The Geometry of Numbers

The Geometry of Numbers

C. D. Olds
Anneli Lax
Giuliana P. Davidoff

Published and Distributed by
The Mathematical Association of America

Dedicated to Dr. Anneli Lax

Editor of the New Mathematical Library
from 1961 to 1991

The Geometry of Numbers is one of the last editorial projects Dr. Lax worked on before her death in October 1999. She observed that we cannot know what Dr. Olds would have added to his book had he lived. Neither do we know how she herself would have organized a final version of the text. I hope that this finished work does justice to the effort and vision of each of them.

—G.P.D.

The New Mathematical Library (NML) was started in 1961 by the School Mathematics Study Group to make available to high school students short expository books on various topics not usually covered in the high school syllabus. In a decade the NML matured into a steadily growing series of some twenty titles of interest not only to the originally intended audience, but to college students and teachers at all levels. Previously published by Random House and L. W. Singer, the NML became a publication series of the Mathematical Association of America (MAA) in 1975. Under the auspices of the MAA the NML continues to grow and remains dedicated to its original and expanded purposes. In its third decade, it contains forty titles.

ANNELI LAX NEW MATHEMATICAL LIBRARY

1. Numbers: Rational and Irrational *by Ivan Niven*
2. What is Calculus About? *by W. W. Sawyer*
3. An Introduction to Inequalities *by E. F. Beckenbach and R. Bellman*
4. Geometric Inequalities *by N. D. Kazarinoff*
5. The Contest Problem Book I Annual High School Mathematics Examinations 1950–1960. Compiled and with solutions *by Charles T. Salkind*
6. The Lore of Large Numbers *by P. J. Davis*
7. Uses of Infinity *by Leo Zippin*
8. Geometric Transformations I *by I. M. Yaglom, translated by A. Shields*
9. Continued Fractions *by Carl D. Olds*
10. Replaced by NML-34
11. } Hungarian Problem Books I and II, Based on the Eötvös Competitions
12. } 1894–1905 and 1906–1928, *translated by E. Rapaport*
13. Episodes from the Early History of Mathematics *by A. Aaboe*
14. Groups and Their Graphs *by E. Grossman and W. Magnus*
15. The Mathematics of Choice *by Ivan Niven*
16. From Pythagoras to Einstein *by K. O. Friedrichs*
17. The Contest Problem Book II Annual High School Mathematics Examinations 1961–1965. Compiled and with solutions *by Charles T. Salkind*
18. First Concepts of Topology *by W. G. Chinn and N. E. Steenrod*
19. Geometry Revisited *by H. S. M. Coxeter and S. L. Greitzer*
20. Invitation to Number Theory *by Oystein Ore*
21. Geometric Transformations II *by I. M. Yaglom, translated by A. Shields*
22. Elementary Cryptanalysis—A Mathematical Approach *by A. Sinkov*
23. Ingenuity in Mathematics *by Ross Honsberger*
24. Geometric Transformations III *by I. M. Yaglom, translated by A. Shenitzer*
25. The Contest Problem Book III Annual High School Mathematics Examinations 1966–1972. Compiled and with solutions *by C. T. Salkind and J. M. Earl*
26. Mathematical Methods in Science *by George Pólya*
27. International Mathematical Olympiads—1959–1977. Compiled and with solutions *by S. L. Greitzer*
28. The Mathematics of Games and Gambling *by Edward W. Packel*
29. The Contest Problem Book IV Annual High School Mathematics Examinations 1973–1982. Compiled and with solutions *by R. A. Artino, A. M. Gaglione, and N. Shell*
30. The Role of Mathematics in Science *by M. M. Schiffer and L. Bowden*
31. International Mathematical Olympiads 1978–1985 and forty supplementary problems. Compiled and with solutions *by Murray S. Klamkin*
32. Riddles of the Sphinx *by Martin Gardner*
33. U.S.A. Mathematical Olympiads 1972–1986. Compiled and with solutions *by Murray S. Klamkin*
34. Graphs and Their Uses *by Oystein Ore*. Revised and updated *by Robin J. Wilson*
35. Exploring Mathematics with Your Computer *by Arthur Engel*
36. Game Theory and Strategy *by Philip D. Straffin, Jr.*

37. Episodes in Nineteenth and Twentieth Century Euclidean Geometry *by Ross Honsberger*

38. The Contest Problem Book V American High School Mathematics Examinations and American Invitational Mathematics Examinations 1983–1988. Compiled and augmented *by George Berzsenyi and Stephen B. Maurer*

39. Over and Over Again *by Gengzhe Chang and Thomas W. Sederberg*

40. The Contest Problem Book VI American High School Mathematics Examinations 1989–1994. Compiled and augmented *by Leo J. Schneider*

41. The Geometry of Numbers *by C.D. Olds, Anneli Lax, and Giuliana Davidoff*

Other titles in preparation.

Books may be ordered from:
MAA Service Center
P. O. Box 91112
Washington, DC 20090-1112
1-800-331-1622   fax: 301-206-9789

Contents

Preface xiii

Part I. Lattice Points and Number Theory

Chapter 1 Lattice Points and Straight Lines

  1.1 The Fundamental Lattice 3
  1.2 Lines in Lattice Systems 4
  1.3 Lines with Rational Slope 6
  1.4 Lines with Irrational Slope 11
  1.5 Broadest Paths without Lattice Points 18
  1.6 Rectangles on Paths without Lattice Points 20
  Problem Set for Chapter 1 23

Chapter 2 Counting Lattice Points

  2.1 The Greatest Integer Function, $[x]$ 25
       Problem Set for Section 2.1 27
  2.2 Positive Integral Solutions of $ax + by = n$ 28
       Problem Set for Section 2.2 31
  2.3 Lattice Points inside a Triangle 32
       Problem Set for Section 2.3 34

Chapter 3 Lattice Points and the Area of Polygons

  3.1 Points and Polygons 37
  3.2 Pick's Theorem 38
       Problem Set for Section 3.2 39
  3.3 A Lattice Point Covering Theorem for Rectangles 40
       Problem Set for Section 3.3 45

Chapter 4 Lattice Points in Circles

4.1 How Many Lattice Points Are There?.....................47
4.2 Sums of Two Squares....................................50
4.3 Numbers Representable as a Sum of Two Squares..........53
   Problem Set for Section 4.3............................56
4.4 Representations of Prime Numbers as Sums of Two Squares. 56
4.5 A Formula for $R(n)$ 58
   Problem Set for Section 4.5............................60

Part II. An Introduction to the Geometry of Numbers

Chapter 5 Minkowski's Fundamental Theorem

5.1 Minkowski's Geometric Approach........................65
   Problem Set for Section 5.1............................66
5.2 Minkowski M-Sets.....................................67
   Problem Set for Section 5.2............................69
5.3 Minkowski's Fundamental Theorem.......................69
   Problem Set for Section 5.3............................74
5.4 (Optional) Minkowski's Theorem in n Dimensions..........75

Chapter 6 Applications of Minkowski's Theorems

6.1 Approximating Real Numbers............................77
6.2 Minkowski's First Theorem.............................78
   Problem Set for Section 6.2............................81
6.3 Minkowski's Second Theorem 81
   Problem for Section 6.3 82
6.4 Approximating Irrational Numbers......................82
6.5 Minkowski's Third Theorem.............................84
6.6 Simultaneous Diophantine Approximations................85
   Reading Assignment for Chapter 6 86

Chapter 7 Linear Transformations and Integral Lattices

7.1 Linear Transformations................................89
   Problem Set for Section 7.1............................92
7.2 The General Lattice...................................92
7.3 Properties of the Fundamental Lattice Λ 94
   Problem Set for Section 7.3............................99
7.4 Visible Points..99

Chapter 8 Geometric Interpretations of Quadratic Forms

   8.1 Quadratic Representation 103
   8.2 An Upper Bound for the Minimum Positive Value 104
   8.3 An Improved Upper Bound 107
   8.4 (Optional) Bounds for the Minima of Quadratic Forms
       in More Than Two Variables 110
   8.5 Approximating by Rational Numbers 111
   8.6 Sums of Four Squares 113

Chapter 9 A New Principle in the Geometry of Numbers

   9.1 Blichfeldt's Theorem 119
   9.2 Proof of Blichfeldt's Theorem 120
   9.3 A Generalization of Blichfeldt's Theorem 121
   9.4 A Return to Minkowski's Theorem 123
   9.5 Applications of Blichfeldt's Theorem 125

Chapter 10 A Minkowski Theorem (Optional)

   10.1 A Brief History of the Question 129
   10.2 A Proof of Minkowski's Theorem 130
   10.3 An Application of Minkowski's Theorem 135
   10.4 Proving the General Theorem 137

Appendix I Gaussian Integers, by Peter D. Lax

   I.1 Complex Numbers 139
       Problem Set for Section I.1 140
   I.2 Factorization of Gaussian Integers 140
       Problem Set for Section I.2 141
   I.3 The Fundamental Theorem of Arithmetic 141
       Problem for Section I.3 144
   I.4 Unique Factorization of Gaussian Integers 144
       Problem for Section I.4 145
   I.5 The Gaussian Primes 145
   I.6 More about Gaussian Primes 148

Appendix II The Closest Packing of Convex Bodies

   II.1 Lattice-Point Packing 151
   II.2 Closest Packing of Circles in R^2 152
   II.3 The Packing of Spheres in R^n 153

Appendix III Brief Biographies
    Hermann Minkowski 157
    Hans Frederik Blichfeldt 159

Solutions and Hints... 161

Bibliography... 169

Index... 172

Preface

The geometry of numbers is a branch of number theory that originated with the publication of Minkowski's seminal work in 1896 and ultimately established itself as an important field of study in its own right. Its focus is the conversion of arithmetic questions into geometric contexts, with the result that certain difficult questions in arithmetic can be answered geometrically by reasonably obvious constructions. One fundamental problem is to define conditions under which a given region contains a lattice point, that is, a point (p, q) with integer coordinates; another is to ask how to characterize regions, $\Re$, for which any point can be moved by an integer translation to coincide with another point in $\Re$, that is, to ask what conditions must hold in $\Re$ if for any point P there is to be a point Q such that $P - Q$ is a lattice point.

Though these questions may seem fairly abstract and detached from practical problems, they are, in fact, crucial to highly significant applications in modern science and technology. Such lattice point problems arise not only in connection with deciding whether certain equations have any integer solutions, but in determining the densest sphere packings or the thinnest sphere coverings of given spaces. Recent extensive research into these latter questions has led to advances in, among other things, crystallography, superstring theory, and the error-detecting and error-correcting codes by which information is stored, compressed for transmission, and received. With the wired and wireless transmission of data and digital information changing the way we live and work, more contemporary or important applications are hard to find.

How can arithmetic problems be restated in geometric terms? A simple example is the question of determining which integers can be represented as sums of two squares; that is, which $n \in \mathbb{Z}$ have the property that there exist integers p and q such that $n = p^2 + q^2$. Alternatively, this same question may be posed as a problem in geometry; namely, we ask whether there are lattice points on the circle $x^2 + y^2 = n$ with center $(0,0)$ and radius $\sqrt{n}$. In a similar fashion, the arithmetic question of whether a linear equation with real coefficients has integer solutions can be transformed into the geometric one asking whether there are lattice points on the corresponding line in the plane.

This latter context suggests many interesting questions beyond the original one: Which lines have lattice points lying arbitrarily close to them? Can two parallel lines bound some strip containing no lattice points?

Part I. Problems like those sketched above motivate the material treated in Chapter 1, while Chapters 2 and 3 extend the investigation to lattice points contained inside or on polygons. Chapter 4 examines various aspects of the circle problem cited above. These four chapters form the first part of the text, which takes as its theme a lattice point approach to number theoretical questions.

Part II. The second part of the text begins the formal introduction to the geometry of numbers. Though almost all definitions and results here can be generalized to n dimensions, this presentation is restricted almost entirely to the more familiar and easily visualized two-dimensional plane. Minkowski's Fundamental Theorem is introduced, motivated, and proved in Chapter 5, while some of his additional theorems are proved or stated in Chapter 6. Minkowski's approach yields extremely good approximations to irrational numbers by rationals; some applications of his method to such "Diophantine approximations" are illustrated in Chapter 6.

Changing course slightly, Chapter 7 provides a discussion of linear transformations that enables the reader to move from the fundamental lattice of integer points to general lattices based on two arbitrary vectors. Taking advantage of that more general context, Chapter 8 goes on to treat one of the central questions in the geometry of numbers, namely, that of determining the minimum value achieved by quadratic forms in two or more variables. Minkowski and, later, Blichfeldt and others gave upper bounds for this minimum, and two applications of their results are presented: first, an improved approximation of irrationals by rationals and, second, a simplified proof of Lagrange's Theorem on the representation of a positive integer as a sum of four squares.

Chapters 9 and 10 are devoted to work of Blichfeldt's that extended Minkowski's approach, provided powerful new insights, and laid the groundwork for many later advances in the geometry of numbers. Chapter 9 introduces some of Blichfeldt's new methods and gives a second proof of Minkowski's Fundamental Theorem that follows from those methods. Chapter 10 gives a proof by Blichfeldt of a theorem of Tchebychev, whose result was later improved by Minkowski. This last theorem arises from what is known as an inhomogeneous problem in the geometry of numbers, and just as Minkowski's approach gives good approximations of irrationals by rationals, so does this one, with the added property that the numerator and denominator can be specified within a given arithmetic progression. The material of these last two chapters will require more work of the reader than the earlier chapters demand—particularly the proof in Chapter 10, which is designated as "optional." However, though the argument is more complex, no mathematical tools beyond some results of solid analytic geometry are required to understand it, and even these may be taken on faith.

The Appendices. When lattice points (p, q) in the plane are identified with the Gaussian integers $p + qi$, they acquire a multiplicative structure in addition to the additive structure explored in the main body of the text. In Appendix I, Peter Lax examines this extra arithmetic structure, giving a lovely geometric proof of the fact that the Gaussian integers form a Euclidean domain, characterizing the Gaussian primes, and proving that unique factorization holds there. In the process, he brings the reader back to the discussion of Chapter 4 and provides yet another glimpse into the power of a geometric approach to number theoretic problems. Appendix II gives a brief survey of the sphere-packing problem and brings the reader up to date with recent results in that area. For the interested reader, Appendix III provides brief biographies of Hermann Minkowski and Hans Frederik Blichfeldt.

Advice to the Reader. C. D. Olds' text and references have been updated and expanded in order to introduce the rich and contemporary subject of the geometry of numbers in a way that is accessible to a broad audience of interested high school students and nonprofessionals. A quick glance at the bibliography shows that many of the seminal works in the field have yet to be translated from the original German. Although excellent introductions to the geometry of numbers exist, they are "advanced" in the sense of requiring the mathematical background of an upper-level undergraduate. As Dr. Olds foresaw when he began this book, a place remains open in the literature for a work of this type, one that allows the interested and motivated reader without a specifically technical background to engage a vitally important topic not addressed by the typical high school or early college curriculum.

Like any text in mathematics, this book should be read with pencil and paper ready. If some step in a proof is not accessible, then return to it later, more than once if necessary, until you have mastered it. Work all the problems given, for this is crucial to testing your understanding of the ideas you will encounter. (Solutions or hints for all but the most obvious are given at the back of the book.) Unfortunately, the elementary problems that can be assigned are few, because only a few steps beyond this book we run into minor—even major—research problems that progressively demand more and more study.

Acknowledgments.   This posthumous volume owes its publication to the unflagging support of Dr. Anneli Lax. Under her guidance the first seven chapters were almost completed and were typed by Gloria Lee; it was Dr. Lax who, with Dr. Olds, designed the figures. Even after Dr. Olds' death, Dr. Lax continued to believe that the book had an important place in the New Mathematical Library series, and her editorial vision assisted me in organizing the various chapters and fragments of text into the form presented here. Dr. Lax herself, along with her husband Peter, continued to proofread the manuscript and offer valuable comments until failing health forced her to limit her editorial activities.

Ellen Curtin did a wonderful job of line-editing and TeX production, not only supplying transitions and adding continuity to rough text, but infusing a tone of accessibility that strongly influenced this finished version. Beverly Ruedi completed the graphics work, interpreting the original sketches with her usual clarity and expertise.

I thank the Department of Mathematics at the University of Rome, La Sapienza, for their generous help in opening their extensive library to me for research on the bibliography. I am grateful to my family and colleagues as well, for their understanding support throughout this long task. Finally, I thank the MAA for offering me the privilege of ushering C. D. Olds' imaginative and still timely exposition into print.

Giuliana P. Davidoff

Part I

Lattice Points and Number Theory

1
Lattice Points
and Straight Lines

1.1 The Fundamental Lattice

The theme of this book is the geometry of numbers, a branch of the theory of numbers that was discovered by Hermann Minkowski (1864–1909). Where other mathematicians had attacked problems of certain types algebraically, Minkowski's genius was to approach them from a geometrical point of view. Through the visible order of geometrical constructs, he was able to reveal and explore many numerical relationships.

We shall trace Minkowski's explorations in the second part of this book. Here, we begin at the beginning, by defining the two elementary concepts on which the entire geometry of numbers rests: the *fundamental lattice L* and the *fundamental point-lattice* Λ that L determines.

When we speak of *lattice systems*, we are imagining grids of points in space connected like monkey bars on a playground. We can construct a simple planar lattice on an ordinary rectangular coordinate system just by drawing straight lines. First we draw lines parallel to the y-axis through the points

$$\ldots, (-2, 0), (-1, 0), (0, 0), (1, 0), (2, 0), (3, 0), \ldots,$$

and then we draw lines parallel to the x-axis through the points

$$\ldots, (0, -2), (0, -1), (0, 0), (0, 1), (0, 2), \ldots.$$

These lines form the fundamental lattice L.

The points where these lines intersect, called *lattice points*, are the vertices of squares. Lattice points are points with integer coordinates (x, y),

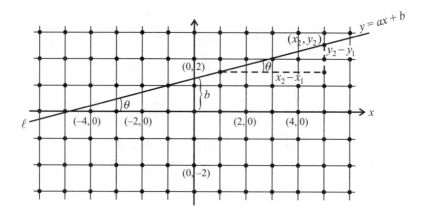

Figure 1.1
The fundamental lattice L, with line ℓ.

as Figure 1.1 indicates. For clarity, to distinguish between a lattice point and any point (x, y), we shall designate a lattice point by (p, q), where $x = p$, $y = q$ are integers.

It is the lattice points that determine the fundamental point-lattice Λ. For now, we shall consider only lattices in the plane, saving n-dimensional figures for Part II.

Evenly spaced along lines running parallel to the x- and y-axes, at first glance lattice points seem to be of little interest in themselves. In 1837, Carl Friedrich Gauss (1777–1855) [1] published a seminal paper on the number of lattice points contained within and on a circle of radius r. Since then, mathematicians have proposed a great many intriguing related problems. Many of these problems have now been answered; some are only partially solved; and others—including the optimal form of Gauss's original question—still defy solution. To prepare ourselves for examining such problems, we must first discuss some of the relationships between straight lines and lattice points.

1.2   Lines in Lattice Systems

On a lattice system, we draw the line

$$\ell : y = ax + b \tag{1.1}$$

having slope a and y-intercept b. When $x = 0$, $y = b$, so the line ℓ passes through the point $(0, b)$ on the y-axis. (See Figure 1.1.) Suppose that (x_1, y_1) and (x_2, y_2) are the coordinates of any two distinct points on ℓ. Then the *slope of ℓ* is

$$a = \tan\theta = \frac{y_2 - y_1}{x_2 - x_1},$$

where θ is the angle ℓ makes with the x-axis, for $x_2 \neq x_1$.

Clearly, if $x_2 = x_1$ the slope a is not defined, for the denominator is zero. Such lines have equations of the form $x = c$, where c is some constant; they are parallel to the y-axis. Likewise, if $y_2 = y_1$, where $x_2 \neq x_1$, then the lines are parallel to the x-axis. Such lines have equations of the form $y = c$, and by definition they have zero slope.

The slope $a = \tan\theta$ of the line ℓ is a real number, hence it must be either *rational* or *irrational*. When the slope a is a rational number, it can be written as the ratio of two relatively prime integers m and n, where $n \neq 0$. We call two integers m and n *relatively prime* if their greatest common divisor is one, denoted by g.c.d.$(m, n) = 1$. Thus, when rational, the slope of ℓ can be written as

$$a = \frac{m}{n}, \quad \text{where g.c.d.}(m, n) = 1 \text{ and } n \neq 0.$$

If the slope a is not rational, then it is an irrational number such as $\sqrt{2}$, $1+\sqrt{3}$, or π. Such numbers cannot be represented as a ratio of two integers. For more details on rational and irrational numbers, Ivan Niven provides an elementary discussion in [**2**] and a deeper discussion in [**3**].

Keeping in mind that a line's slope can be rational or irrational, we ask the question:

Through how many lattice points does the line $y = ax + b$ pass?

The answer turns out to depend largely on the slope. As we shall see, every such line belongs to one of five types:

1. Lines with rational slope containing no lattice points.
2. Lines with rational slope containing an infinite number of lattice points.
3. Lines with irrational slope containing no lattice points.
4. Lines with irrational slope containing exactly one lattice point.

5. Lines parallel to the y-axis (of the form $x = k$) with undefined slope containing either infinitely many lattice points or none, depending on whether the constant k is an integer or not.

Note that lines parallel to the x-axis (of the form $y = k$) belong to either the first or the second type. Lattice point considerations relative to lines $x = k$ and $y = k$ are so obvious that we shall largely ignore them from now on.

Think about these different cases, and you will begin to see interesting questions. For example, suppose a line contains infinitely many lattice points: How are they distributed along that line? Or, suppose a line passes through no lattice points—or only one: Can we find lattice points arbitrarily close to it? These are some of the questions we will be answering as we discuss the five types of lines in lattice systems.

1.3 Lines with Rational Slope

The first two types of lines in lattice systems that we shall explore are those with rational slope, which have either no lattice points or infinitely many lattice points. Consider a line

$$y = \frac{m}{n}x + b, \quad \text{where g.c.d.}(m,n) = 1 \text{ and } n \neq 0, \qquad (1.2)$$

having rational slope $a = m/n$. Recall that g.c.d.$(m,n) = 1$ means that m and n are relatively prime integers. A natural question to ask is:

What are the necessary and sufficient conditions that line (1.2) *will pass through a lattice point* $(x,y) = (p,q)$?

First of all, suppose the lattice point (p,q) is on the line (1.2). Then p and q satisfy the equation

$$q = \frac{m}{n}p + b.$$

Multiplying by $n \neq 0$ gives us $nq = mp + nb$, so

$$b = \frac{nq - mp}{n}.$$

Since b is a ratio of two integers, b is a rational number. Hence, we can write b in the form $b = r/s$, where $(r,s) = 1$ and $s \neq 0$. Line (1.2) now has the form

$$y = \frac{m}{n}x + \frac{r}{s}, \quad \text{where g.c.d.}(m,n) = \text{g.c.d.}(r,s) = 1. \qquad (1.3)$$

Thus, we see that any line with rational slope passing through a lattice point must have rational intercept, that is, must have the form (1.3).

Now, let's look at the sufficient conditions:

Does every line of the form (1.3) *pass through a lattice point?*

Consider, for instance, the line

$$y = \frac{3}{4}x - \frac{1}{5}, \tag{1.4}$$

which is of this type, since $(3, 4) = 1$ and $(1, 5) = 1$. If $(x, y) = (p, q)$ were a lattice point on line (1.4), then the coordinates (p, q) would satisfy the equation

$$q = \frac{3}{4}p - \frac{1}{5}.$$

Thus, $5(3p - 4q) = 4$. But this implies that 4 is divisible by 5, which is false. Hence, there are no lattice points on the line (1.4). This shows that not every line of the form (1.3) passes through a lattice point. Can we decide which ones do?

Let's state the desired conditions more precisely. To do so, we shall need two symbols.

1. The notation $s|n$ means that s *is a factor of* n or that n *is exactly divisible by* s. This relation implies that $n = ks$, where k is some integer.
2. The notation $s \nmid n$ means that s *does not divide* n.

We will need several results about divisibility properties of integers as we proceed, so we introduce those results now.

Preliminary Lemma. *Suppose that m and n are integers with greatest common divisor d_0, that is, g.c.d.$(m, n) = d_0$. Then there exist integers s_0 and t_0 such that $ms_0 + nt_0 = d_0$.*

Proof. Consider the set $S = \{ms + nt = d | s, t \in \mathbb{Z}\}$. When $t = 0$ and $s = \pm 1$, we see that $|m| \in S$, so S contains at least one positive integer and, hence, a non-empty subset of positive integers, S'. Let $d = ms_0 + nt_0$ be the smallest integer in S'. There may be many pairs of integers s and t that give the same value of d; in fact, there are. For now, we pick one such pair and let it contain the s_0 and t_0 we consider. We now claim that this smallest d in S' is the greatest common divisor of m and n; in other words, $d_0 = d$.

First, we must show that if some other integer d_1 divides both m and n, then $d_1 | d$. But this is clear, since if $m = d_1 k$ and $n = d_1 r$, then

$$d = d_1 k s_0 + d_1 r t_0$$
$$= d_1 (k s_0 + r t_0)$$
$$= d_1 k'.$$

We must show that $d | m$ and $d | n$. Now, given m and d, we know that there exist $q, r \in \mathbb{Z}$ such that

$$m = dq + r$$
$$= (m s_0 + n t_0) q + r, \qquad \text{where } 0 \leq r < d.$$

But this means that we can write

$$r = m - (m s_0 + n t_0) q$$
$$= m (1 - s_0 q) + n (-t_0)$$
$$= m s_1 + n t_1,$$

showing that if $r \neq 0$, then $r \in S'$. However, the latter cannot be true, since then S' would contain a positive integer smaller than d, contradicting our choice of d. Hence, we must have $r = 0$ and $d | m$. An identical argument shows that $d | n$, proving that $d = d_0$.

The Fundamental Theorem of Arithmetic. *If an integer c divides the product ab of two integers a and b, and if* g.c.d.$(c, a) = 1$*, then* $c | b$*.*

Proof. Since the least common multiple l.c.m.$(c, a) = 1$, we must have integers s and t such that $cs + at = 1$. Multiplying both sides by b, we get $cbs + abt = b$. But we know that $c | cbs$ and $c | abt$. Therefore $c | b$, and the proof is complete.

Now, if (p, q) is a lattice point on the line (1.3), then

$$q = \frac{m}{n} p + \frac{r}{s}, \quad \text{where g.c.d.}(m, n) = \text{l.c.m.}(r, s) = 1,$$

which simplifies to

$$s(nq - mp) = nr.$$

That means $s | nr$, which reduces to $s | n$ because g.c.d.$(r, s) = 1$.

We have shown that *if there is a lattice point* (p, q) *on line* (1.2), *then b is a rational number of the form* $b = r/s$, *where* $(r, s) = 1$ *and where* $s|n$. Hence, it is now quite simple to exhibit equations of type (1), with rational slope and containing no lattice points. We need only write equations of the form (1.3) and make sure that $s \nmid n$. Equation (1.4) gives one such line.

Now for lines of type (2). We will show that if there is *one* lattice point $(x, y) = (p_0, q_0)$ on the line

$$y = \frac{m}{n}x + \frac{r}{s}, \quad \text{where g.c.d.}(m, n) = \text{g.c.d.}(r, s) = 1 \text{ and } s|n, \quad (1.5)$$

then the equations

$$\begin{aligned} p_k &= p_0 + kn \\ q_k &= q_0 + km, \end{aligned} \quad \text{for } k = 0, \pm 1, \pm 2, \dots,$$

produce not only (p_0, q_0)—which we get by taking $k = 0$—but also an infinite number of other lattice points (p_k, q_k) on line (1.5) as k assumes, in turn, the values $k = \pm 1, \pm 2, \dots$.

This statement is proved as follows. Suppose by some process we have established that (p_0, q_0) is actually a lattice point on the line (1.5). Then

$$q_0 = \frac{m}{n}p_0 + \frac{r}{s}, \quad \text{where g.c.d.}(m, n) = \text{g.c.d.}(r, s) = 1. \quad (1.6)$$

We do not yet know whether line (1.5) has other lattice points. But for the moment we shall assume that it has another lattice point (p, q), so that

$$q = \frac{m}{n}p + \frac{r}{s}. \quad (1.7)$$

By subtracting (1.6) from (1.7), we get the relation

$$n(q - q_0) = m(p - p_0), \quad (1.8)$$

which implies that $n|m(p - p_0)$. However, g.c.d.$(m, n) = 1$, so $n \nmid m$, and we are forced to conclude that $n|(p - p_0)$. This means that $p - p_0 = kn$, or $p = p_0 + kn$, where k is some integer. Substituting $p - p_0 = kn$ into (1.8) gives

$$n(q - q_0) = m(kn),$$

from which we easily conclude that

$$q = q_0 + mk.$$

So far, we have demonstrated that if (p_0, q_0) is any lattice point on line (1.5), and if there happens to be another lattice point $(p, q) = (p_k, q_k)$ on the same line, then

$$p = p_k = p_0 + kn,$$
$$q = q_k = q_0 + km, \tag{1.9}$$

for some integer k. Because the values of p and q depend on the integer k, we have switched the notation from (p, q) to (p_k, q_k).

Now it is easy to verify the converse: *If (p_0, q_0) is a lattice point on line* (1.5), *then equations* (1.9) *with $k = 0, \pm 1, \pm 2, \ldots$, automatically produce infinitely many additional lattice points on this line.* To show this, we simply substitute $x = p_k, y = q_k$ from (1.9) into (1.5) and get

$$q_0 + km = \frac{m}{n}(p_0 + kn) + \frac{r}{s},$$

or

$$q_0 \frac{m}{n} p_0 + \frac{r}{s} + (mk - mk),$$

which, for $k = 0, \pm 1, \pm 2, \ldots$, obviously reduces to

$$q_0 = \frac{m}{n} p_0 + \frac{r}{s}.$$

This is a true identity, since we assumed that (p_0, q_0) was a lattice point on line (1.5).

Remarks. Our discussion so far has been keyed to the phrase: "If (p_0, q_0) is a lattice point on (1.5), then...." But are we sure that such a lattice point actually exists? Under the conditions stated in (1.5), it can be proved by means of Euclid's algorithm [**5**], or by continued fractions [**4**, pp. 46–48], that such a solution (p_0, q_0) always exists. Sometimes a good guess will produce this first solution (p_0, q_0) of (1.5).

This concludes our proof that a line with rational slope contains infinitely many lattice points if and only if the conditions in (1.5) are satisfied. In particular, if the last condition is violated—that is, if $s \nmid n$—then the line with rational slope contains no lattice point.

1.4 Lines with Irrational Slope

We shall now examine types (3) and (4) from our list relating straight lines to lattice points. We want to prove that the line

$$y = ax + b \qquad (1.10)$$

with irrational slope a either contains no lattice points or passes through one, and only one, lattice point.

Let's consider the alternative:

What if there were two distinct lattice points (p_1, q_1) and (p_2, q_2) on line (1.10)?

Then by the slope formula, we would have

$$a = \tan \theta = \frac{q_2 - q_1}{p_2 - p_1},$$

where $q_2 - q_1$ and $p_2 - p_1 \neq 0$ are both integers; so the slope a would be rational, contrary to the hypothesis. Thus such a line must have either one lattice point or none on it.

It is easy to write equations of lines that illustrate these two possibilities. For example, the line $y - q = a(x - p)$, where a is irrational, passes through any given lattice point (p, q); but no other lattice points are on this line. Clearly, the line $x = c$ where c is not an integer, contains no lattice points.

To be more general, let's choose any *rational but not integral* constant b in equation (1.10). Then

$$y = ax + b, \;\; \text{so} \;\; b = y - ax.$$

If this line had a lattice point $(x, y) = (p, q)$, where $x = p \neq 0$, then $b = q - ap$ would be irrational, since a is irrational and p and q are integers. But b was assumed to be rational. If $x = 0$, there could be no lattice point on this line, for $(x, y) = (0, q)$ would imply that $b = q$ is an integer. Hence, a line with irrational slope and rational, non-integral intercept b contains no lattice point.

Remarks. The equation of a line with irrational slope raises an interesting question: For what values of θ will the slope $a = \tan \theta$ be irrational? It can be proved [3] that $\tan \theta$ is irrational if θ, measured in radians, is any nonzero rational number. Moreover, $\tan \theta$ is irrational if θ, measured in degrees, is any nonzero rational number except $45 + 90n$, where $n = 0, \pm 1, \pm 2, \ldots$.

A line with irrational slope, then, cannot pass through more than one lattice point. Either it misses them all, or it misses all but one. The question of interest for us is:

By what distance do lines with irrational slope miss lattice points?

The answer is given by the following theorem, which says that although a line can avoid all lattice points, it comes arbitrarily close to an infinite number of them. By this we mean that given any distance, no matter how small, an infinite number of lattice points will lie even closer than that to the line. Formally, we may state this as follows.

Theorem 1.1. *Any line $y = ax + b$, where a is irrational and where b is any real number, has on either side an infinite number of lattice points lying closer to it than any assigned distance $\epsilon > 0$, no matter how small a value is chosen for ϵ.*

For the proof of Theorem 1.1, we need a preliminary theorem, or *lemma*.

Lemma 1.1. *Let a be any irrational number, let c be any positive real number, and let $\epsilon > 0$ be any arbitrarily small number. Then we can always find a pair of integers (p_1, q_1) such that*

$$c < p_1 a - q_1 < c + \epsilon; \qquad (1.11)$$

that is,

$$0 < p_1 a - q_1 - c < \epsilon. \qquad (1.11')$$

Similarly, we can find a pair of integers (p_2, q_2) such that

$$c - \epsilon < p_2 a - q_2 < c; \qquad (1.12)$$

that is,

$$-\epsilon < p_2 a - q_2 - c < 0. \qquad (1.12')$$

Proof of Theorem 1.1. Referring to Figure 1.2, let's suppose that $\epsilon_1 > 0$ is given. If $b < 0$ and Lemma 1.1 is true, then, letting $c = -b$, we can

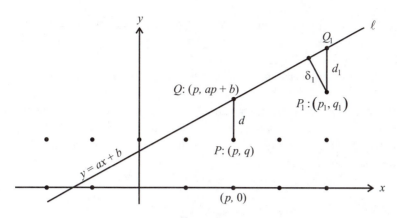

Figure 1.2

Finding nearby lattice points.

find a lattice point $P_1 : (p_1, q_1)$ such that $0 < ap_1 - q_1 + b < \epsilon_1$. Thus,

$$0 < d_1 = y - q_1 = ap_1 + b - q_1 < \epsilon_1$$

and, because $d_1 > 0$, the point P_1 lies below the line.

The actual distance between P_1 and ℓ is the perpendicular distance δ_1, which is less than d_1. Hence, $0 < \delta_1 < \epsilon_1$. On the other hand, if $b > 0$, we find an $n \in \mathbb{Z}^+$ such that $b - n < 0$, and we let $c = -(b - n)$. Then, again, Lemma 1.1 ensures that we can find $P_1 : (p_1, q_1)$ such that

$$0 < ap_1 - q_1 + (b - n) < \epsilon.$$

Thus,

$$0 < d_1 = ap_1 + b - (q_1 + n)$$
$$= y - (q_1 + n) < \epsilon_1.$$

So, again $P_1 : (p_1, q_1 + n)$ lies below the line and δ_1 satisfies $0 < \delta_1 < \epsilon_1$.

Our lemma implies that for every ϵ there is a pair of integers p, q satisfying inequality (1.11′). So by picking a sequence of decreasing positive numbers $\epsilon_1 > \epsilon_2 > \cdots > 0$, we can construct an infinite sequence of lattice points $P_1, P_2, P_3, \ldots$, all lying below ℓ, with each P_i within perpendicular distance ϵ_i of ℓ.

To exhibit the presence of infinitely many lattice points above the line ℓ, we repeat the argument. In this case, however, the points lie above ℓ, so $q > y = ap + b$. Therefore, $ap - q + b < 0$, so we must use inequality (1.12'). This completes the proof of Theorem 1.1, except for the lemma.

Note. The proof of Lemma 1.1 can be omitted at first reading.

Proof of Lemma 1.1. It is enough to prove the lemma for the special case where $a = \alpha$ and $c = \gamma$ are positive numbers between 0 and 1. To see this, suppose that a were any irrational number and c any real number. Then we could rewrite each as follows.

First, since a is irrational, we can write it in the form

$$a = [a] + \alpha, \quad \text{where } 0 < \alpha < 1.$$

The function denoted by $[x]$ is defined for every real number x as *the largest integer not exceeding x*; this important function is discussed further in Chapter 2.

Likewise, since c is real, we can write it in the form

$$c = [c] + \gamma, \quad \text{where } 0 \leq \gamma < 1.$$

Therefore, if we were able to find integers p_1 and q such that

$$\gamma < p_1 \alpha - q$$
$$< \gamma + \epsilon,$$

we could substitute $\alpha = a - [a]$ and $\gamma = c - [c]$, thus arriving at

$$c - [c] < p_1 \big(a - [a] \big) - q$$
$$< c - [c] + \epsilon.$$

Adding the integer $[c]$ to both sides of these inequalities gives

$$c < p_1 a - p_1 [a] - q + [c]$$
$$< c + \epsilon.$$

We now give the name q_1 to the integer $p_1[a] + q - [c]$ and obtain the desired inequality (1.11):

$$c < p_1 a - q_1$$
$$< c + \epsilon.$$

Now we prove the lemma for irrational α, where $0 < \alpha < 1$, and for a number γ, with $0 \leq \gamma < 1$. We draw a circle Γ of radius $1/(2\pi)$ (hence of circumference 1) and select a point A on its circumference. Starting at A

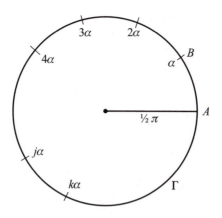

Figure 1.3

Integer multiples of length α marked on Γ.

and moving along the circumference, we mark out lengths $\alpha, 2\alpha, 3\alpha, \ldots$. In Figure 1.3, for example, the arc AB has length α. Now, suppose we come to the point $j\alpha$, where j is some integer, having started from A and gone around the circle Γ a certain integer number of times—say, $n_1 = [j\alpha]$ times—plus some extra distance α_1, where $0 < \alpha_1 < 1$. We can write our new point as

$$j\alpha = n_1 + \alpha_1, \quad \text{for } 0 < \alpha_1 < 1, \tag{1.13}$$

where j and n_1 are integers. Similarly, any other marked point $k\alpha$, where $k \neq j$ is an integer, can be expressed in the form

$$k\alpha = n_2 + \alpha_2, \quad \text{for } 0 < \alpha_2 < 1, \tag{1.14}$$

where $n_2 = [k\alpha]$ is an integer. Since α is irrational, no two points $j\alpha$ and $k\alpha$ can coincide. For if they did, then $\alpha_1 = \alpha_2$; and subtracting (1.14) from (1.13) would yield $(j - k)\alpha = n_1 - n_2$, or

$$\alpha = \frac{n_1 - n_2}{j - k}, \quad \text{where } j \neq k,$$

making α rational, contrary to the hypothesis.

What does our intuition tell us? As we go around and around Γ, plotting in succession the infinite number of points $\alpha, 2\alpha, 3\alpha, \ldots$, no two of them can coincide. So in the neighborhood of at least one point on Γ, an infinite number of these points must "cluster." Call one such cluster point Q. Then, no matter how small an arc we designate on the circumference

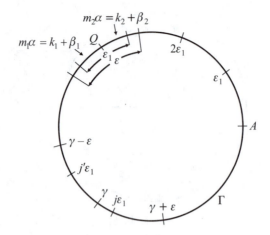

Figure 1.4

Q, γ, and the points $n\epsilon_1$ on Γ.

around Q, we will always be able to find an infinite number of points $n\alpha$ on that arc.

Remarks. If you prefer, this intuitive argument can be made completely rigorous by means of an important theorem of analysis called the Bolzano-Weierstrass Theorem. This theorem states that *every bounded infinite set of points has at least one limit or cluster point.* For discussion and proof, consult any advanced calculus textbook.

Again, we let this cluster point on the circle Γ be the point Q, as in Figure 1.4. Among the points that cluster about Q, there must be at least two, say

$$m_1\alpha = k_1 + \beta_1,$$
$$m_2\alpha = k_2 + \beta_2, \qquad \text{where } \beta_1 \neq \beta_2,$$

whose distance apart, measured on Γ, is less than a given arc length $\epsilon > 0$, no matter how small ϵ may be. In other words, the distance between them, measured along the circumference, is some number ϵ_1, where $0 < \epsilon_1 < \epsilon$. Let's arrange the subscripts so that $\beta_1 > \beta_2$; now we can describe their separation as

$$\epsilon_1 = \beta_1 - \beta_2 = (m_1\alpha - k_1) - (m_2\alpha - k_2) = m\alpha - k, \qquad (1.15)$$

where $m = m_1 - m_2$ and $k = k_1 - k_2$ are integers.

Let's pause a moment and get our bearings. On Γ we have marked off an infinite set of points at distance $n\alpha$ from A, where $n = 1, 2, 3, \ldots$. These points cluster around some point Q. We have set some small arc length $\epsilon > 0$ around Q and have found two points $m_1\alpha$ and $m_2\alpha$ on the resulting arc whose distance apart is $\epsilon_1 < \epsilon$. We can now disregard the set of points $n\alpha$ and begin a second construction.

To proceed, starting again from A on a clean copy of our circle Γ, we mark the point whose distance from A, measured along the circumference, is our second number γ, where $0 \leq \gamma \leq 1$. (Recall that γ is the number that is replacing c in our statement of Lemma 1.1.) We also mark off the points at distance $\epsilon_1, 2\epsilon_1, 3\epsilon_1, \ldots$ from A. Then there will be a first j such that γ will lie on the arc between $(j-1)\epsilon_1$ and $j\epsilon_1$. That is,

$$(j-1)\epsilon_1 \leq \gamma < j\epsilon_1. \tag{1.16}$$

The distance between any two consecutive points of this new set is $\epsilon_1 < \epsilon$. Hence in Figure 1.4, the arc from the point $\gamma - \epsilon$ to the point γ, and the arc from γ to $\gamma + \epsilon$, each contain at least one point of the set $n\epsilon_1$.

We know that the latter arc contains $j\epsilon_1$ because of the determination (1.16) of j. The former arc contains the point $(j-1)\epsilon_1$ if γ is not an integer multiple of ϵ_1; otherwise it contains the point $(j-2)\epsilon_1$. Thus we can always find integers j' and j such that

$$\gamma < j\epsilon_1 < \gamma + \epsilon,$$
$$\gamma - \epsilon < j'\epsilon_1 < \gamma.$$

Moreover, since from (1.15) we have $\epsilon_1 = m\alpha - k$, we can find integers j' and j such that

$$\gamma < jm\alpha - jk < \gamma + \epsilon,$$
$$\gamma - \epsilon < j'm\alpha - j'k < \gamma.$$

This means that we have have found integers $p_1 = jm$, $q_1 = jk$, $p_2 = j'm$, and $q_2 = j'k$ such that

$$\gamma < p_1\alpha - q_1 < \gamma + \epsilon,$$
$$\gamma - \epsilon < p_2\alpha - q_2 < \gamma.$$

This completes the proof of Lemma 1.1.

1.5 Broadest Paths without Lattice Points

We have seen that there are lines that contain no lattice points. The question then arises:

Are there infinite strips, or paths, between parallel lines that are also lattice point–free?

Theorem 1.1 tells us that the answer is no if the parallel lines have irrational slope. However, in the case of strips defined by lines with rational slope, the answer to this question is yes. In this section we shall look for the broadest lattice point–free paths associated to such lines.

Through the origin $(0,0)$ of the fundamental point-lattice Λ, we draw a line ℓ' at an angle θ with the x-axis. Without loss of generality, we can assume that $0 \leq \theta \leq \pi/2$ (for, if $\pi/2 < \theta < \pi$, we would be working with the mirror image of ℓ' in the y-axis and our results would be the same). Next, we draw a second line ℓ'' above (or below) ℓ' and parallel to it at a distance d, as in Figure 1.5.

The region between ℓ' and ℓ'' is called a *path* of width d in the direction θ. Clearly, in the two cases where $\theta = 0$ and $\theta = \pi/2$, this path is of width $d = 1$. Our question is:

What is the broadest path in the direction θ that contains no lattice points in its interior?

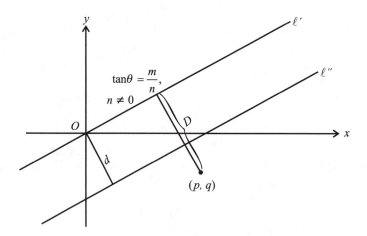

Figure 1.5

A path of width d.

When θ is such that $\tan \theta$ is irrational, then we know by Theorem 1.1 that every path of finite width d in direction θ contains infinitely many lattice points in its interior.

What if $\tan \theta$ is rational, however? (That is, $\tan \theta = m/n$, where m and $n \neq 0$ are relatively prime integers.) In this case there are paths of width $w > 0$ containing no lattice points. The following theorem is a precise description of this situation.

Theorem 1.2. *If* $\tan \theta = m/n$, *where* g.c.d.$(m, n) = 1$ *and* $n \neq 0$, *then there is a path of width* $d = 1/\sqrt{m^2 + n^2}$ *in the direction* θ *that does not contain lattice points in its interior; furthermore, every path of width greater than d contains lattice points.*

Proof. The perpendicular distance D of any lattice point (p, q) from ℓ' is given by an analytic geometry formula,

$$D = \frac{|mp - nq|}{\sqrt{m^2 + n^2}}. \tag{1.17}$$

(This standard formula is derived by writing the equation of the line $y = (m/n)x$ in the "normal form":

$$\frac{(mx - ny)}{\sqrt{m^2 + n^2}} = 0.$$

For any point (p, q) in the plane, the expression

$$\frac{|mp - nq|}{\sqrt{m^2 + n^2}}$$

gives the distance from (p, q) to ℓ'.)

Our theorem will be proved if we can show that the distance from ℓ' to the nearest lattice point not actually on ℓ' is equal to $1/\sqrt{m^2 + n^2}$.

First we examine the numerator in (1.17). Because g.c.d.$(m, n) = 1$, we can always find two integers $p = p_1$ and $q = q_1$ and, hence, a lattice point (p_1, q_1) such that $mp_1 - nq_1 = \pm 1$ [**4**, pp. 36–42]. Symbolically, we can write this as

$$|mp_1 - nq_1| = 1.$$

This is the smallest nonzero value possible for the numerator in formula (1.17), for if the numerator in D were zero, then (p, q) would be a lattice point on ℓ'.

The denominator in (1.17) has the value $\sqrt{m^2 + n^2}$, which is fixed once the slope of ℓ' is given. Thus the distance from ℓ' to the nearest lattice point (p_1, q_1) is given by

$$d = \frac{1}{\sqrt{m^2 + n^2}}.$$

Consider the width w of the path between ℓ' and ℓ''. If their distance apart is d or less, the path has no lattice points; but if $w > d$, it has. Thus, $d = 1/\sqrt{m^2 + n^2}$ determines the *path of maximum width* in the direction θ that is free from interior lattice points, and Theorem 1.2 is proved.

1.6   Rectangles on Paths without Lattice Points

It is interesting to find the area of a rectangle formed when ℓ' and ℓ'' are the borders of a lattice point–free path of maximum width.

Suppose one border—say, ℓ'—passes through the origin $O : (0,0)$ and through a second lattice point $C_1 : (p, q)$, where $(p, q) = 1$. As we see in Figure 1.6, a rectangle can be formed by $OA_1B_1C_1$, where

$$|\overline{OA_1}| = |\overline{C_1B_1}| = d$$

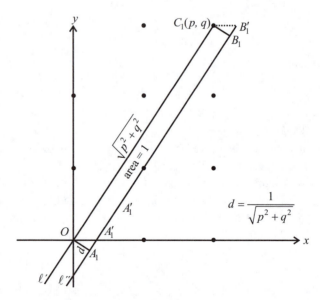

Figure 1.6

Unit rectangle in a lattice point–free path.

and

$$d = \frac{1}{\sqrt{p^2 + q^2}}.$$

This rectangle has *unit area*.

The proof is nearly obvious. We see that length $|\overline{OC_1}| = \sqrt{p^2 + q^2}$; we know that width $d = 1/\sqrt{p^2 + q^2}$; and the area of the rectangle is equal to their product.

This result on unit area can also be derived geometrically, using a special case. Notice first in Figure 1.6 that the area of the rectangle $OA_1B_1C_1$ is equal to the area of the parallelogram $OA_1'B_1'C_1$. Now compare this to Figure 1.7, where ℓ' is a line of slope $\tan\theta = m/n = 3/2$, for g.c.d.$(3,2) = 1$. This line passes through $O : (0,0)$ and through the lattice point $C : (2,3)$.

What lattice point is nearest to ℓ'? It is the point $(1,1)$, for by the distance formula (1.17), the distance from $(1,1)$ to ℓ' is equal to

$$D = \frac{|3 \cdot 1 - 2 \cdot 1|}{\sqrt{3^2 + 2^2}} = \frac{1}{\sqrt{13}}.$$

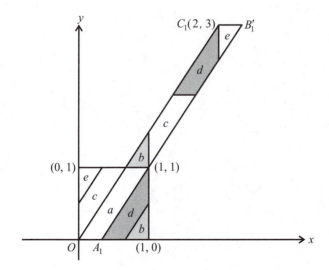

Figure 1.7

Forming a unit square.

Hence, if we draw the line ℓ'' through $(1,1)$ parallel to ℓ', no lattice points can be interior to the parallelogram $OA_1'B_1'C_1$, whose area is made up of the parts labeled a through e.

Now we will rearrange the parts in Figure 1.7 to build a unit square. Leaving part a where it is, we slide parts b and c downward, parallel to the y-axis. Leaving part b in its new position near $(1,0)$, we continue sliding part c parallel to the x-axis to its new position, at the left of part a. Similarly, we slide parts d and e down, slipping part d to the right of part a and part e to the left of c. The unit square is now completely filled, which is what we set out to show.

The rectangle $OA_1B_1C_1$ of area 1 in Figure 1.6 is expanded in Figure 1.8. The expanded rectangle $B_1B_2B_3B_4$ is made up of four rectangles, all congruent to $OA_1B_1C_1$. Thus it has area 4. It is also *symmetric with respect to the origin*; this phrase means that, for every point (a,b) belonging to a figure, the point $(-a,-b)$ also belongs to the figure.

There are no lattice points inside $OA_1B_1C_1$; nor, since $(p,q) = 1$ are there any lattice points between $(0,0)$ and (p,q). Hence, the origin $(0,0)$ is the only lattice point inside $B_1B_2B_3B_4$. At least two lattice points are

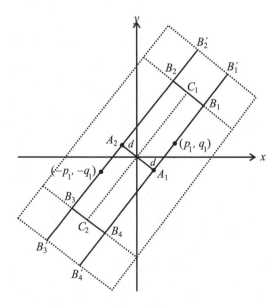

Figure 1.8
Expanding the rectangle.

on the boundary, namely $C_1 : (p, q)$ and $C_2 : (-p, -q)$. But there must also be a lattice point (p_1, q_1) somewhere on side $A_1 B_1$ and, by symmetry, another lattice point $(-p_1, -q_1)$ on side $B_3 A_2$. So we have at least four lattice points on the boundary of $B_1 B_2 B_3 B_4$.

What happens if we expand this rectangle, preserving the symmetry? For example, let's step beyond sides $B_1 B_2$ and $B_3 B_4$ to $B_1' B_2'$ and $B_3' B_4'$, as indicated by the dotted lines in Figure 1.8. The area of our new rectangle will be greater than 4, and its interior will contain at least two lattice points—namely (p, q) and $(-p, -q)$—besides $(0, 0)$. In fact, as long as we preserve symmetry, no matter how we stretch our rectangle by moving a pair of opposite sides to make a new rectangle of area greater than 4, the new boundary will always contain at least two lattice points besides $(0, 0)$.

This discussion suggests that *any rectangle symmetric about the origin $(0, 0)$ and of area greater than 4 encloses at least two lattice points other than the origin $(0, 0)$.*

Immediately, other questions come to mind. Would such a statement hold for an ellipse centered at $(0, 0)$? How general is this result? Suppose a rectangle of sides a by b is placed anywhere on the fundamental point-lattice Λ: What dimensions of a and b will ensure that this rectangle contains at least one lattice point within it or on its boundary? Some questions like these can be approached at an elementary level, as we shall see in subsequent chapters.

Problem Set for Chapter 1

Hints for easy problems and the complete solutions of the more difficult ones are given at the end of this book.

1. Give three examples of lines with rational slope containing

   a. no lattice points;

   b. an infinite number of lattice points.

2. Prove that the lattice points (p_k, q_k) given by equations (1.9) for some integer $k = 0, \pm 1, \pm 2, \ldots$, are equally spaced on the line (1.5).

3. A line $y = mx + b$ passes through the lattice points (p_1, q_1) and (p_2, q_2). Prove that it also passes through the lattice points (p_k, q_k), where $p_k = p_1 + k(p_2 - p_1)$ and $q_k = q_1 + k(q_2 - q_1)$, for k any integer.

4. A line $y = (m/n)x$, where m and n are relatively prime integers, passes through a lattice point (p, q), where $(p, q) = 1$. Prove that there are no other lattice points on this line between $(0, 0)$ and (p, q).

5. Consider the line $y = \sqrt{2}x$. Does it pass through any lattice points? Explain why or why not.

6. For each $\epsilon > 0$, find lattice points (p, q) whose distance from the line $y = \sqrt{2}x$ is less than ϵ:

   a. $\epsilon = \frac{1}{2}$

   b. $\epsilon = \frac{1}{5}$

   c. $\epsilon = \frac{1}{10}$

7. Prove that every triangle with three non-collinear vertices at lattice points, but with no other lattice points on its boundary and no interior lattice points, has an area of $\frac{1}{2}$.

References

1. C. F. Gauss, *Werke*, Vol. 2 (Göttingen: Gesellschaft der Wissenschaften, 1876).

2. Ivan Niven, *Numbers: Rational and Irrational*, New Mathematical Library Series, Vol. 1 (New York and Toronto: Random House, 1961).

3. ———, "Simple Irrationalities," Chapter 2, Section 2, in *Irrational Numbers*, Carus Mathematical Monographs, No. 11 (New York: Wiley, 1956), 16–21.

4. Carl D. Olds, *Continued Fractions*, New Mathematical Library Series, Vol. 9 (New York and Toronto: Random House, 1963).

5. J. V. Uspensky and M. A. Heaslet, *Elementary Number Theory* (New York: McGraw-Hill, 1939).

2

Counting Lattice Points

2.1   The Greatest Integer Function, [x]

Frequently, we will wonder how many lattice points occur on line segments, or inside rectangles, or in various parts of conic sections, and so on. Basically, what we want to know is: *How do we count lattice points, or at least estimate their number?* This chapter offers some ideas.

We shall again make use of the arithmetical function $[x]$, defined for every real number x as *the largest integer not exceeding x*:

$$[x] = \text{largest integer} \le x.$$

We call this integer the *integral part of x*. For example, the integral part of 3.6, denoted by $[3.6]$, is 3, because 3 is the largest integer less than or equal to 3.6. Similarly, $[6] = 6$, $[-2] = -2$, and $[-2.5] = -3$. The equation $y = [x]$ is graphed in Figure 2.1.

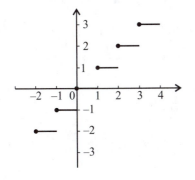

Figure 2.1
Graph of $y = [x]$.

Figure 2.2

The fractional part of x.

Thinking of the position of $[x]$ on the number line, how might we express its relation to neighboring integers? By its definition, the integer $[x]$ satisfies the inequalities $[x] \leq x < [x] + 1$. Consequently, we can write

$$x = [x] + \zeta, \quad \text{where } 0 \leq \zeta < 1.$$

We call ζ the *fractional part of* x; it is indicated in Figure 2.2.

The function $[x]$ has many useful properties, four of which are listed here. Try testing each one with numerical examples.

1. $[x + n] = [x] + n$, if n is an integer.

2. $[x] + [-x] = 0$, if x is an integer;
   $\qquad\qquad = -1$, if x is not an integer.

3. $[x] + [y] \leq [x + y] \leq [x] + [y] + 1$.

4. $\left[\frac{[x]}{n}\right] = \left[\frac{x}{n}\right]$, if n is a positive integer.

The following proof of property (3) illustrates the basic ideas for proving the rest of the properties.

Proof of Property 3. We shall prove that $[x + y] \geq [x] + [y]$. Let

$$x = [x] + \zeta_1, \quad \text{where } 0 \leq \zeta_1 < 1, \quad \text{and}$$

$$y = [y] + \zeta_2, \quad \text{where } 0 \leq \zeta_2 < 1.$$

Their sum is

$$x + y = [x] + [y] + (\zeta_1 + \zeta_2).$$

Hence,

$$[x + y] = \left[(\zeta_1 + \zeta_2) + ([x] + [y])\right].$$

Because $[x] + [y]$ is an integer, we can invoke property (1), writing

$$[x + y] = [\zeta_1 + \zeta_2] + [x] + [y].$$

Now, our conditions were $0 \le \zeta_1 < 1$ and $0 \le \zeta_2 < 1$, so $0 \le \zeta_1 + \zeta_2 < 2$. Therefore, depending on whether $0 \le \zeta_1 + \zeta_2 < 1$ or $1 \le \zeta_1 + \zeta_2 < 2$, we can evaluate $[\zeta_1 + \zeta_2]$ as being either 0 or 1.

Correspondingly,

$$[x + y] = \begin{cases} [x] + [y] & \text{if } [\zeta_1 + \zeta_2] = 0, \\ [x] + [y] + 1 & \text{if } [\zeta_1 + \zeta_2] = 1, \end{cases}$$

so we have proved that $[x + y] \ge [x] + [y]$, as stated in property (3).

Problem Set for Section 2.1

1. Give counterexamples to show that none of the these statements is true for all x and y.

   a. $[x + y] = [x] + [y]$.

   b. $\left[\frac{x}{y}\right] = \frac{[x]}{[y]}$.

   c. $[xy] = [x] \cdot [y]$.

2. Prove property (1).

3. Prove property (2).

4. Attempt to prove property (4) for $[x]$, first studying a numerical example. (Don't be discouraged if you have to look at the solution in the back of the book.)

5. Prove that $[2x] + [2y] \ge [x] + [y] + [x + y]$.

6. Prove that if a and b are positive integers, then the number of positive multiples of b less than or equal to a is $[a/b]$.

7. Prove that $-[-x]$ is the least integer greater than or equal to x.

8. Prove that $\left[x + \frac{1}{2}\right]$ is the nearest integer to x. If x is midway between two integers, $\left[x + \frac{1}{2}\right]$ gives the larger of the two.

9. As in problem 8, make and prove similar statements about $-\left[-x + \frac{1}{2}\right]$.

10. The following result is proved in all number theory textbooks: If n is a positive integer, the exponent of the highest power of a prime p that divides $n! = 1 \cdot 2 \cdot 3 \cdot \cdots \cdot n$ is

$$E(p, n) = \left[\frac{n}{p}\right] + \left[\frac{n}{p^2}\right] + \left[\frac{n}{p^3}\right] + \cdots,$$

a sum that clearly has only finitely many nonzero terms. Use this formula to show that $E(7, 1000) = 164$; that is, 7^{164} is the highest power of 7 in $1000!$.

2.2 Positive Integral Solutions of $ax + by = n$

Let the equation of a line ℓ have the form

$$\ell : ax + by = n \quad \text{for g.c.d.}(a, b) = 1, \tag{2.1}$$

where a, b, and n are positive integers. Notice that we assume a and b to be relatively prime. We are interested in regarding this equation as a *Diophantine equation*; that is, its coefficients are integers and we seek integer solutions x and y. We have shown in Chapter 1 that the equation $ax + by = n$ in (2.1) has an infinite number of such solutions $x = p_k$, $y = q_k$, where p_k and q_k are both integers. This implies that the line (2.1) passes through an infinite number of lattice points (p_k, q_k) of Λ. Suppose (p_0, q_0) is one of its lattice points. Then all the other lattice points (p_k, q_k) on ℓ can be calculated from the equations

$$\begin{aligned} p_k &= p_0 + kb, \\ q_k &= q_0 - ka, \end{aligned} \quad \text{where } k = 0, \pm 1, \pm 2, \ldots. \tag{2.2}$$

(You may recognize this type of equation; we discussed others of the same form in Section 1.3, although we did not require the coefficients in equation (1.5) to be positive.)

The line ℓ in (2.1) intersects the x-axis at the point $A : (n/a, 0)$ and the y-axis at $B : (0, n/b)$. This pattern suggests an interesting question:

Is there a formula for finding the number of lattice points on ℓ between the points A and B?

In other words, how many positive integral solutions (x, y) does equation (2.1) have?

We approach this question by examining Figure 2.3, where the right triangle OAB has legs of lengths n/a and n/b. The formula for the length c of the hypotenuse $\overline{AB}$ tells us

$$c = \sqrt{\left(\frac{n}{a}\right)^2 + \left(\frac{n}{b}\right)^2} = \frac{n}{ab}\sqrt{a^2 + b^2}.$$

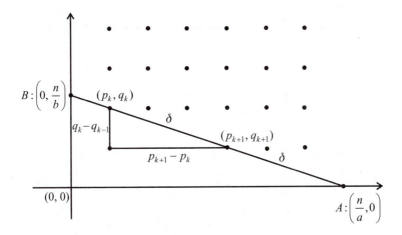

Figure 2.3

The distance between consecutive lattice points on ℓ.

Replacing k by $k+1$ in (2.2) to indicate the next lattice point, we see that

$$p_{k+1} = p_0 + (k+1)b = p_k + b,$$

$$q_{k+1} = q_0 - (k+1)a = q_k - a.$$

From here we determine the distance δ between any two consecutive lattice points $P_k : (p_k, q_k)$ and $P_{k+1} : (p_{k+1}, q_{k+1})$ on ℓ to be

$$\delta = \sqrt{(p_{k+1} - p_k)^2 + (q_{k+1} - q_k)^2} = \sqrt{a^2 + b^2}.$$

We are ready to estimate the number N of lattice points on $\overline{AB}$. First, we divide $c = |\overline{AB}|$ by $\delta = |\overline{P_k P_{k+1}}|$, to obtain $c/\delta = n/ab$. Then

$$\left[\frac{c}{\delta}\right] = \left[\frac{n}{ab}\right]$$

gives the number of complete δ-intervals we can mark off on $\overline{AB}$. Finally, since the endpoints of each interval are lattice points, we can estimate how many lattice points are on $\overline{AB}$.

There is a complication, however. Because k adjacent δ-intervals on $\overline{AB}$ are separated by $k - 1$ subdivision lattice points, and because c/δ might not be an integer, we must consider the following separate cases. Each of the three cases is depicted in Figure 2.4.

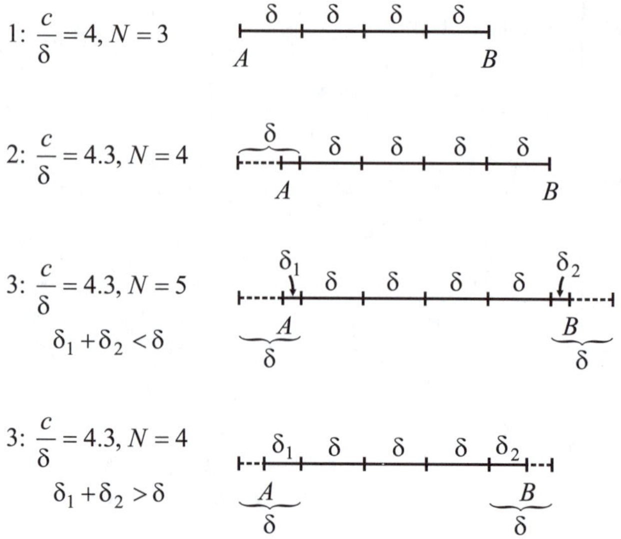

Figure 2.4

Whole and fractional δ-intervals. Cases 1, 2, and 3.

Case 1. Both intercepts A and B are lattice points.

This implies that both n/b and n/a are integers, or that $a|n$ and $b|n$, where g.c.d.$(a, b) = 1$. By the Fundamental Theorem of Arithmetic in Chapter 1 it follows that $ab|n$. (However, the two conditions $a|n$ and $b|n$ alone do not necessarily imply that $ab|n$; for instance, $3|12$, $6|12$, but $18 \nmid 12$.)

In this case, we have precisely $N = (n/ab) - 1 = [n/ab] - 1$ lattice points on $\overline{AB}$, not counting endpoints A and B. Hence, the number of positive integral solutions (x, y) of equation (2.1) on $\overline{AB}$ is $[n/ab] - 1$.

But if we allow x and y to have zero values, then the number of nonnegative solutions (x, y) of equation (2.1) on $\overline{AB}$ is $[n/ab] - 1 + 2 = [n/ab] + 1 = N + 2$.

Case 2. Only one intercept, A or B, is a lattice point.

In this case, n is divisible by either a or b, but not both, so n/ab is not an integer. For example, say that B is a lattice point (so $b|n$) while A is not (so $a \nmid n$). Then A falls inside a δ-interval that will extend beyond A; see Figure 2.4. Excluding B, the number of lattice points between A and B is the number of complete δ-intervals there. This number is $[n/ab]$, and it represents the number N of positive integral solutions of (2.1) on $\overline{AB}$. There will be $[n/ab] + 1 = N + 1$ nonnegative solutions.

Case 3. Neither intercept A nor intercept B is a lattice point.

Here both endpoints A and B are inside δ-intervals that extend in two directions beyond A and B, as illustrated in Figure 2.4. In this case, the number of lattice points on $\overline{AB}$ is one greater than the number of δ-intervals on $\overline{AB}$.

We see this by summing the lengths of the *fractional parts* δ_1 and δ_2 between A and B. Two outcomes are possible. First, if $\delta_1 + \delta_2 < \delta$, then the number of whole δ-intervals is $[n/ab]$, and the number N of lattice points on $\overline{AB}$ is one greater: $[n/ab] + 1$. Second, if $\delta_1 + \delta_2 > \delta$, then the number of whole δ-intervals is $[n/ab] - 1$, and the number of lattice points is still one greater: $N = [n/ab]$ lattice points on $\overline{AB}$.

These cases establish the following theorem for counting lattice points.

Theorem 2.1. *If a, b, and n are positive integers, with g.c.d.$(a, b) = 1$, then the number of lattice points $(x, y) = (p, q)$, where $x = p > 0$ and $y = q > 0$, on the line $ax + by = n$ is equal to*

$$N = \left[\frac{n}{ab}\right] + \zeta, \quad \text{for } ab \neq 0, \tag{2.3}$$

where ζ has one of the values -1, 0, or 1.

Although formula (2.3) is not a precise formula for N, it does produce three consecutive integers one of which is equal to N. We saw that the case $\zeta = -1$ can occur only when n/ab is an integer. So, clearly, equation (2.1) always has at least one positive solution (x, y) if $n > ab$. Dickson [1] offers more information along the lines of formula (2.3).

Problem Set for Section 2.2

1. Verify the arguments leading to Theorem 2.1 using the following numerical examples. Draw figures for each case.

   a. $x + y = 5$.   d. $3x + 2y = 13$.

   b. $2x + y = 5$.   e. $4x + 3y = 11$.

   c. $3x + 4y = 24$.

2. Prove that if (p_0, q_0) is any particular solution of (2.1), then the equations for (p_k, q_k) given in (2.2) provide all other solutions.

3. Illustrate Theorem 2.1 using a numerical example.

4. Howard Grossman offered many interesting problems in his "Fun with Lattice Points" [2]. William Schaaf compiled a listing of his publications on the topic, as well as an extensive bibliography of other problem books in geometry [3]. Here is one example of the type of problem the interested reader will find:

The number of lattice paths from the origin $O : (0,0)$ to lattice points on the line $x + 2y = n$ is equal to the nth term of the Fibonacci sequence $1, 2, 3, 5, 8, 13, 21, \ldots$ defined by $u_1 = 1, u_2 = 2, u_3 = u_1 + u_2 = 3, \ldots, u_n = u_{n-1} + u_{n-2}$. A lattice path is like the route of a taxi allowed to travel from O only north (up) or east (to the right) along the lattice. This means, in particular, that only lattice points with nonnegative integer coordinates are accessible. Show by a drawing that if $n = 7$, the number of paths from O to the lattice points $(7, 0)$, $(5, 1)$, $(3, 2)$, $(1, 4)$ is $1 + 6 + 10 + 4 = 21 = u_7$.

2.3 Lattice Points inside a Triangle

How do you find the sum of a series of integers? For certain series, this problem can often be solved by considering it geometrically in terms of lattice points. Here is a typical example.

Theorem 2.2. *If P and Q are two positive, relatively prime integers,* $(P, Q) = 1$, *then*

$$\left[\frac{Q}{P}\right] + \left[\frac{2Q}{P}\right] + \left[\frac{3Q}{P}\right] + \cdots + \left[\frac{(P-1)Q}{P}\right] = \frac{(Q-1)(P-1)}{2}.$$

Proof. On the fundamental lattice Λ plot the points $O : (0,0)$, $A : (P, 0)$, $B : (P, Q)$, and $C : (0, Q)$; see Figure 2.5, which is drawn for $P = 7, Q = 5$. Without loss of generality, we can assume that $P > Q$, for if $P < Q$ we can simply swap their labels.

Because the equation of the diagonal OB is

$$y = \frac{Q}{P}x, \quad \text{where g.c.d.}(Q, P) = 1, \tag{2.4}$$

the only lattice points on OB are O and B. For if there were a lattice point (p, q) on line (2.4) between O and B, then $q/p = Q/P$. Yet $0 < p < P$

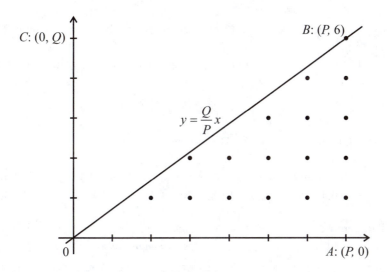

Figure 2.5

Lattice points in a triangle.

and $0 < q < Q$, which contradicts the fact, guaranteed by the hypothesis g.c.d.$(Q, P) = 1$, that the fraction Q/P is in lowest terms.

Now, letting $x = n$ in line (2.4), where $n = 1, 2, \ldots, P - 1$, we have

$$[y] = \left[\frac{nQ}{P}\right].$$

This integer represents the y-coordinate of the lattice point that is on the vertical line $x = n$ closest to the line's point (n, y), and below the line. From $[y]$ we also obtain a count of the number of lattice points on $x = n$ below OB and above the x-axis. Hence we can now calculate how many lattice points are inside $\triangle OAB$.

The total number of lattice points in the interior of $\triangle OAB$ is equal to the sum

$$\left[\frac{1Q}{P}\right] + \left[\frac{2Q}{P}\right] + \cdots + \left[\frac{(P-1)Q}{P}\right]. \tag{2.5}$$

For example, in Figure 2.5 the total number of lattice points inside $\triangle OAB$ on the lines $x = 1, 2, 3, 4, 5$, and 6 is equal to

$$\left[\frac{1 \cdot 5}{7}\right] + \left[\frac{2 \cdot 5}{7}\right] + \left[\frac{3 \cdot 5}{7}\right] + \left[\frac{4 \cdot 5}{7}\right] + \left[\frac{5 \cdot 5}{7}\right] + \left[\frac{6 \cdot 5}{7}\right] = 12.$$

By symmetry, the rectangle $OABC$ contains exactly twice this sum: 24 lattice points.

In general, then, the sum (2.5) for the triangle is exactly half the total number of lattice points

$$\frac{(P-1)(Q-1)}{2}$$

inside the rectangle $OABC$. This proves Theorem 2.2.

Many results are known that extend and generalize Theorem 2.2. Here are two additional theorems; their proofs are exercises in the problem set.

Theorem 2.3. *If P and Q are two positive integers, and if $d = $ g.c.d.(P, Q), where g.c.d.(P, Q) denotes the greatest common divisor of P and Q, then*

$$\left[\frac{1 \cdot Q}{P}\right] + \left[\frac{2 \cdot Q}{P}\right] + \cdots + \left[\frac{(P-1)Q}{P}\right] = \frac{(P-1)(Q-1)}{2} + \frac{d-1}{2}.$$

Theorem 2.4. *If $P' = (P-1)/2$, and $Q' = (Q-1)/2$, where P and Q are odd primes, then*

$$\sum_{j=1}^{P'} \left[\frac{jQ}{P}\right] + \sum_{j=1}^{Q'} \left[\frac{jP}{Q}\right] = P'Q'.$$

Problem Set for Section 2.3

1. Explain why the right side $\frac{1}{2}(P-1)(Q-1) + \frac{1}{2}(d-1)$ of the formula in Theorem 2.2 is an integer for all choices of integers P, Q with greatest common divisor d.
2. Invent numerical examples illustrating Theorems 2.2, 2.3, and 2.4.
3. Prove Theorem 2.3.
4. Prove Theorem 2.4.

References

1. L. E. Dickson, "Linear Diophantine Equations and Congruences," Chapter 2 in *History of the Theory of Numbers, Vol. II: Diophantine Analysis* (Washington, D.C.: Carnegie Institute, 1920), 64–71.

2. Howard D. Grossman, "Fun with Lattice Points," *Scripta Mathematica* 16 (1950): 207–12.
3. William Schaaf, *Bibliography of Recreational Mathematics*, Vol. I (Reston, VA: National Council of Teachers of Mathematics, 1959; reprinted, 1973).

3
Lattice Points and the Area of Polygons

3.1 Points and Polygons

Many interesting relationships exist between lattice points and the areas of geometrical figures such as polygons and rectangles. Later, in Part II, we will be exploring Minkowski's beautiful theorems on this fascinating *geometry of numbers*. This chapter introduces the basic concepts underlying these relationships. We will begin by defining key terms, then examine two important theorems.

By a *polygon*, we mean a set of points called *vertices* connected in a given order by line segments called *sides*; see Figure 3.1. To construct a polygon, we number the given points $P_1, P_2, \ldots, P_n$, then draw segments $\overline{P_1 P_2}, \overline{P_2 P_3}, \ldots, \overline{P_{n-1} P_n}, \overline{P_n P_1}$. Two consecutive sides $P_{k-1}P_k$ and $P_k P_{k+1}$ have the vertex P_k in common. In a *simple polygon*, no two sides have any other points in common. The two polygons at the left and

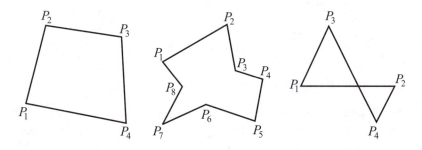

Figure 3.1

Left, center: Examples of simple polygons. *Right:* A non-simple polygon.

37

center of Figure 3.1 are simple polygons; the one on the right is not. We shall be concerned only with simple polygons.

The *boundary* of a polygon is the set of all its sides and vertices. The boundary divides the plane into two regions, namely, the polygon's *interior* and its *exterior*. As we look for lattice points, we will be checking locations within, on, and outside of these boundaries.

It is easy to test whether a point P is in a polygon's interior:

1. First, we enclose the entire polygon in a large circle C having no points in common with the polygon; see Figure 3.2. Since there are only a finite number n of vertices $P_1, P_2, \ldots, P_n$, a circle enclosing all n of them can always be drawn.

2. Next we draw an arbitrary ray from P in such a direction that it avoids all n vertices, and we continue until the ray intersects C—let's say, in a point Q.

3. Now we count how many times this ray PQ crosses the boundary of the polygon. If it crosses an odd number of times, then P is in the interior; if PQ crosses the boundary an even number of times, P is exterior to the polygon. These possibilities are represented by P and P' in Figure 3.2, with rays intersecting C at Q and Q', respectively.

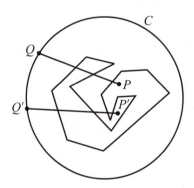

Figure 3.2
Determining whether P is on the interior or exterior of a polygon.

3.2   Pick's Theorem

Let $\mathcal{P}$ be any simple polygon with vertices on our lattice of integers, Λ. An interesting relationship exists between the area of $\mathcal{P}$ and the number of lattice points contained on and in it. According to Hugo Steinhaus, this relationship was first proved by Georg Pick in 1899; it is appropriately called *Pick's Theorem*.

Theorem 3.1. *The area of any simple polygon $\mathcal{P}$ whose vertices are lattice points of Λ is given by the formula*

$$A = I + \frac{1}{2}B - 1, \qquad (3.1)$$

where I is the number of lattice points inside $\mathcal{P}$ and B is the number of lattice points on the boundary of $\mathcal{P}$, including vertices.

The usual proof of Pick's Theorem is based on this fact: A triangle whose vertices are lattice points, but which has no other lattice points either in its interior or on its sides, has area precisely $\frac{1}{2}$. Pick's Theorem has been adequately discussed elsewhere [1], so we omit its proof.

Problem Set for Section 3.2

1. Plot the rectangle with vertices $(0, 0), (a, 0), (a, b), (0, b)$, where a and b are integers. Calculate, in terms of a and b, the number of lattice points B (on the boundary) and I (interior), thus verifying Pick's Theorem for this rectangle.

2. Plot the triangle $(0, 0), (a, 0), (a, b)$, where a and b are positive relatively prime integers. Use Pick's Theorem to show that the number of lattice points interior to this triangle for is equal to $I = \frac{1}{2}(a-1)(b-1)$.

3. Plot the polygon $P_1 P_2 P_3 P_4$, where $P_1 = (0, 0)$, $P_2 = (6, 3)$, $P_3 = (3, 0)$, and $P_4 = (1, 2)$. Show that Pick's Theorem does not hold for $P_1 P_2 P_3 P_4$. Is this a simple polygon?

4. Plot the polygon $\mathcal{P} : (0, 0), (6, 0), (6, 2), (4, 5), (1, 3)$. Omit the inner polygon $\mathcal{P}_1 : (2, 1), (4, 2), (4, 3), (2, 2)$. Does Pick's Theorem hold for the doubly connected polygon $\mathcal{P} - \mathcal{P}_1$?

5. Consider any polygon $\mathcal{P}$ and a second polygon $\mathcal{P}'$ whose boundary lies entirely inside $\mathcal{P}$. Supposes all vertices of $\mathcal{P}$ and $\mathcal{P}'$ are lattice points. Denote by $\mathcal{P} - \mathcal{P}'$ the region bounded by the boundaries of $\mathcal{P}$ and $\mathcal{P}'$. Show that Pick's Theorem does not hold for the doubly connected polygon $\mathcal{P} - \mathcal{P}'$ and that the formula gives an area that is one square unit less than the true area.

6. Consider the rhombus $a|x| + b|y| = ab$, where a and b are positive integers with a greatest common divisor g.c.d.$d = (a, b)$. Find a formula for the number of lattice points interior to the rhombus.

3.3   A Lattice Point Covering Theorem for Rectangles

Remember that Λ is our fundamental lattice of integer points in the xy-plane. A rectangle is said to have the *lattice point covering property* if it always has at least one lattice point of Λ in its interior, or on its boundary, regardless of where it is placed in the plane. You might guess that the rectangle's size determines whether it has this property, and indeed the following theorem spells out the size criterion.

Theorem 3.2.   *Any rectangle with dimensions a and b (say, with $a \leq b$) has the lattice point covering property if and only if $a \geq 1$ and $b \geq \sqrt{2}$.*

Proof.   The conditions stated in the theorem are necessary. For consider the alternative that $a < 1$ and b is any size we please. Then we could align the rectangle with side b parallel to the y-axis, so that no lattice points were inside it or on its boundary; see rectangle R_1 in Figure 3.3. Such a rectangle obviously does not have the lattice point covering property. Similarly, we cannot have a arbitrary and $b < 1$.

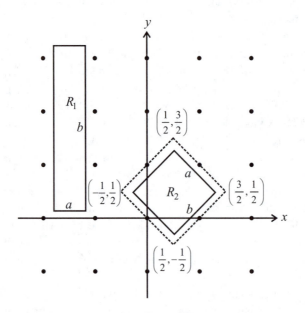

Figure 3.3

Rectangles R_1, with $a < 1$, and R_2, with $1 \leq a \leq b < \sqrt{2}$.

On the other hand, if $a \geq 1$ but $b < \sqrt{2}$ (that is, $1 \leq a \leq b < \sqrt{2}$), we can tilt the rectangle at $45°$ with the x-axis so that it lies completely inside the square having vertices at $(\frac{1}{2}, -\frac{1}{2}), (\frac{3}{2}, \frac{1}{2}), (\frac{1}{2}, \frac{3}{2}), (-\frac{1}{2}, \frac{1}{2})$ and sides of length $\sqrt{2}$; see rectangle R_2 in Figure 3.3. Again, a rectangle so placed lacks the covering property. Hence if every rectangle of dimensions a and b is to have the lattice point covering property, we certainly must require $a \geq 1$ and $b \geq \sqrt{2}$.

To prove the converse, we need the following lemma.

Lemma 3.1. *Let ℓ_1 and ℓ_2 be two parallel lines a distance $\sqrt{2}$ apart. The strip consisting of these two lines and the space between them contains, in either direction, infinitely many lattice points.*

Proof of Lemma 3.1. Suppose the parallel lines ℓ_1 and ℓ_2, a distance of $\sqrt{2}$ apart, are parallel to either the x- or y-axis; see Figure 3.4, upper left.

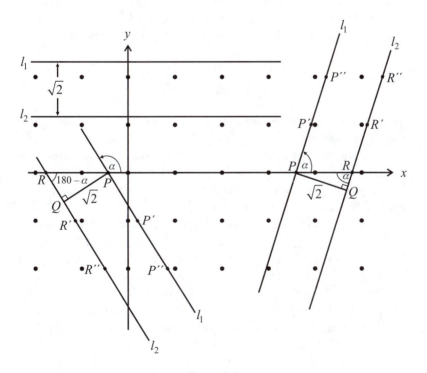

Figure 3.4
Lattice points on paths of width $\sqrt{2}$.

Then a line of the lattice L either lies between ℓ_1 and ℓ_2 or coincides with one of them, and such a line contains infinitely many lattice points of Λ. So in such cases the lemma is true.

On the other hand, suppose the lines ℓ_1 and ℓ_2 intersect the x-axis at an angle α, where $0 < \alpha < 90°$ or $90° < \alpha < 180°$, as at the bottom and right of Figure 3.4. Then $0 < \sin \alpha < 1$, and from the right triangle PQR we see that, for the line segment $\overline{PR}$,

$$|\overline{PR}| = \frac{\sqrt{2}}{\sin \alpha} > \sqrt{2} > 1.$$

So, having width greater than $\sqrt{2}$, $\overline{PR}$ must contain at least one lattice point. But the same holds for the parallel spans $\overline{P'R'}, \overline{P''R''}, \ldots$, where the parallel lines ℓ_1 and ℓ_2 cut the horizontal lines of the lattice L. Hence, any strip formed by ℓ_1 and ℓ_2 contains infinitely many lattice points. This proves Lemma 3.1.

Proof of Theorem 3.2 (continued). It suffices to prove that a rectangle of dimensions 1 by $\sqrt{2}$ has the lattice point covering property. For let us suppose the contrary. Then we can somehow place a rectangle $ABCD$, where

$$|\overline{AB}| = |\overline{CD}| = \sqrt{2} \quad \text{and}$$
$$|\overline{BC}| = |\overline{AD}| = 1$$

on the lattice L so that this rectangle has no lattice points of Λ either inside it or on its boundary.

Clearly, the sides of this rectangle cannot be parallel to either the x- or y-axis; otherwise, at least one lattice point would be inside or on $ABCD$. Instead we may presume that side AB of the rectangle does not parallel either axis; see Figure 3.5a.

Extending sides AD and BC in either direction, we form a strip of width $\sqrt{2}$, which by Lemma 3.1 contains infinitely many lattice points. Now, if we slide $ABCD$ up or down this strip, side AB (or CD) must therefore strike or pass through infinitely many lattice points.

Consider the position of this rectangle when it first encounters a lattice point P on AB (or Q on CD). Let's call this shifted position $A'B'C'D'$ and assume that P on AB is the point in question. Look at Figure 3.5b. Where will we find P? As far as we know, it could be at A', at B', or anywhere along the segment connecting them.

Because P is the first lattice point encountered by the displaced rectangle, it must also be the only lattice point of Λ inside or on $A'B'C'D'$.

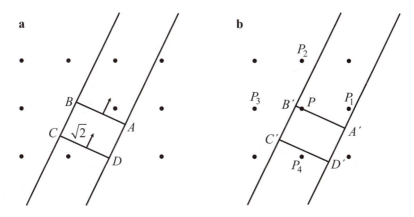

Figure 3.5

(a) A rectangle on a path of width $\sqrt{2}$. (b) Sliding the rectangle to the first lattice point P.

For suppose another point R lay on or inside the displaced rectangle, as in Figure 3.5b. Then either side AB must have encountered R before P while sliding to its new position $A'B'C'D'$ (contrary to our assumption that P comes first), or else $ABCD$ contained R in the first place (contrary to our assumption that $ABCD$ originally has no lattice points of Λ). Thus the situation in Figure 3.5b will not occur if $ABCD$ lacks the covering point property and if P is determined by our method.

But now we will demonstrate that if $ABCD$ has sides of length 1 and $\sqrt{2}$, as assumed, then the situation of Figure 3.5b must occur. In other words, at least one of the lattice points closest to P must lie in the displaced rectangle $A'B'C'D'$. In Figure 3.5a, these closest lattice points are denoted P_1, P_2, P_3, and P_4.

To show this, we draw a circle Γ of radius 1 and center P as in Figure 3.6. The diameter through A' and B' cuts the diameter through P_1 and P_3, since $A'B'$ is not horizontal. Hence, two of P's lattice neighbors—say, P_3 and P_4—lie below the line through A' and B'. These two lattice points also lie above the line through C' and D', because $C'D'$ is the tangent line to Γ at Q; hence, $C'D'$ lies below any point on the circumference, when viewed from Q. Thus we have shown that P_3 and P_4 lie within the strip determined by $A'B'$ and $C'D'$.

We now wish to show that either P_3 or P_4 must lie between the edges $B'C'$ and $A'D'$ or on one of them. To do so, we denote the angle QPP_4

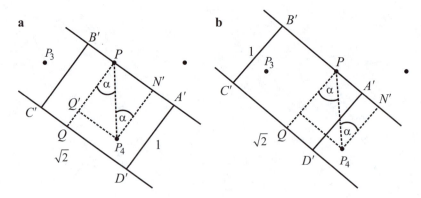

Figure 3.6

A circle Γ with center P.

by α and use it to examine the position of P_4, which, in our construction, lies to the right of the line through $B'C'$. Let one perpendicular from P_4 intersect the line through $A'B'$ at a point N', while a second intersects $\overline{PQ}$ at Q'. If $|PA'| \geq \sin\alpha$, then this tells us that

$$|\overline{PA'}| \geq \frac{|\overline{PN'}|}{|\overline{PP_4}|} = |PN'| = \sin\alpha.$$

Hence,

$$|\overline{PA'}| \geq |\overline{Q'P_4}|$$

and, in that case, P_4 must be to the left of the side $A'D'$ or on it.

On the other hand, suppose

$$|\overline{PA'}| < \sin\alpha = |\overline{PN'}|.$$

In that case, P_4 lies outside $A'B'C'D'$. But if so, the distance from P_4 to $B'C'$ must be greater than $|\overline{A'B'}| = \sqrt{2}$; consequently, a circle with center P_4 and radius $\sqrt{2}$ does not intersect $B'C'$ yet goes through P_3. So if P_4 lies outside the rectangle, then P_3 must lie inside it.

The assumption that ABCD can be positioned on the lattice L in such a way as to contain no lattice points thus leads to a contradiction, and Theorem 3.2 is proved.

This proof of Theorem 3.2 is due to I. Niven and H. Zuckerman [2], who also developed other interesting covering theorems.

Problem Set for Section 3.3

1. Show by experiment (such as careful drawings) that the number $\sqrt{2}$ in Lemma 3.1 can be replaced by a smaller number. How much smaller? Will this smaller number give a covering theorem?

2. Accurately reproduce the rightmost strip in Figure 3.4, and on clear plastic cut out an accurate $\sqrt{2} \times 1$ rectangle. Convince yourself that, if there is a lattice point P on $A'B'$ and a lattice point P_4 inside $A'B'C'D'$, then there is also a lattice point inside $ABCD$.

3. Mark the point where the diagonals of the transparent rectangle in Problem 2 intersect. Place this on a lattice point.

   a. Will this rectangle always have another lattice point in it or on it?

   b. How would you change the dimensions to make this happen?

   c. How much larger could these dimensions be?

4. Make a transparent 2×2 square and mark the point where its diagonals intersect. Use this to answer the questions in Problem 3.

References

1. Ross Honsberger, *Ingenuity in Mathematics*, New Mathematical Library Series, Vol. 23 (New York: Random House, 1970), 27–31.
2. Ivan Niven and Herbert Zuckerman, "The Lattice Point Covering Theorem for Rectangles," *Mathematics Magazine* 42 (1969):85–86.

4
Lattice Points in Circles

4.1 How Many Lattice Points Are There?

Among the earliest explorations of lattice points were those undertaken by
C. F. Gauss. In 1837 Gauss [3] published a result addressing the question
of how many lattice points occur within or on a circle of a certain size.
Using our terminology, we would phrase his question this way:

*What is the number $N(n)$ of lattice points in the interior and on the
boundary of a circle $C(\sqrt{n})$ that has radius $r = \sqrt{n}$ and is centered
at the origin of the fundamental point-lattice Λ, with n a nonnegative
integer?*

Gauss calculated the numerical results from 10 to 300 presented in Table
4.1.

Table 4.1. The Number, $N(n)$, of Lattice Points for Circles of Radius r

$r = \sqrt{n}$	$N(n)$	$N(n)/n$	$r = \sqrt{n}$	$N(n)$	$N(n)/n$
1	5	5	9	253	$\vdots$
2	13	3.25	10	317	3.17
3	29	$3.\overline{22}$	20	1257	3.1425
4	49	3.0625	30	2821	$3.13\overline{4}$
5	81	3.24	100	31417	3.1417
6	113	$\vdots$	200	125629	3.140725
7	149	$\vdots$	300	282697	$3.1410\overline{7}$
8	197	$\vdots$			

The evidence in the table suggests that as n increases, the ratio of $N(n)$ to n gets closer and closer to π. Symbolically, we would like to write

$$\lim_{n \to \infty} \frac{N(n)}{n} = \pi = 3.14159\ldots.$$

In 1961 Mitchell [6] calculated additional values of $N(n)$, from $\sqrt{n} = 1$ to $\sqrt{n} = 200,000$. He took big jumps in the values after $\sqrt{n} = 1000$, quite understandably, as he was working on what is now considered an extremely slow computer. A few of his calculations are excerpted in Table 4.2.

Table 4.2. Lattice Point Count for Larger r

$r = \sqrt{n}$	$N(n)$	$N(n)/n \approx \pi$
400	502 625	3.14141...
500	785 349	3.14139...
1 000	3 141 549	3.141549...
10 000	314 159 221	3.141592...
100 000	31 415 939 281	3.141594...
200 000	125 663 759 077	3.141594...

Again, as with Gauss's results, the numerical evidence is strongly suggestive. Table 4.2 gives added support to the conjecture that the ratio $N(n)/n$ tends to π as n tends to ∞. Stated symbolically, this would mean that the quantity

$$\left| \frac{N(n)}{n} - \pi \right| \tag{4.1}$$

becomes as small as we please provided n is taken large enough.

Geometry can help us understand the behavior of the ratio $N(n)/n$ more clearly. Multiplying (4.1) by n, we obtain the new quantity

$$\left| N(n) - n\pi \right|,$$

which contains a familiar component. We recognize $n\pi$ as the area of a circle of radius $r = \sqrt{n}$. This clue tells us how to approach Gauss's question geometrically.

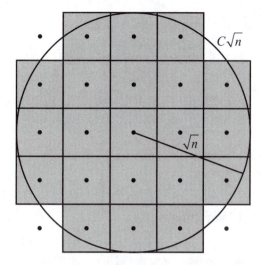

Figure 4.1
Unit squares centered in and on circle $C(\sqrt{n})$.

We want to estimate $N(n)$, the number of lattice points in and on a circle $C(\sqrt{n})$ of arbitrary radius $r = \sqrt{n}$. First, we draw such a circle and center it at the origin $(0,0)$ of the fundamental point-lattice Λ. Now, we let each lattice point of Λ be the center of a unit square with sides parallel to the coordinate axes, and we shade all the squares whose centers are inside or on $C(\sqrt{n})$, as in Figure 4.1. This shaded area is equal to $N(n)$.

Observe, however, that some parts of the shaded region are outside of the disk $x^2 + y^2 \le n$, and that the disk is not entirely shaded. This observation lets us bound the shaded area $N(n)$ from below and above; we simply find the largest disk whose interior is completely shaded, and the smallest disk whose exterior is completely unshaded. Because the diagonal of the unit square is $\sqrt{2}$, all the shaded squares must be contained in a circle of radius $r = \sqrt{n} + (\sqrt{2}/2)$. Similarly, the circle of radius $r = \sqrt{n} - (\sqrt{2}/2)$ is contained entirely within the shaded squares.

It follows that

$$\pi \left(n - \sqrt{2n} - \frac{1}{2} \right) \le \pi \left(n - \sqrt{2n} + \frac{1}{2} \right) = \pi \left(\sqrt{n} - \frac{\sqrt{2}}{2} \right)^2$$

$$\le N(n) \le \pi \left(\sqrt{n} + \frac{\sqrt{2}}{2} \right)^2 = \pi \left(n + \sqrt{2n} + \frac{1}{2} \right).$$

This relation implies that

$$\left| \frac{N(n)}{n} - \pi \right| \leq \pi \left(\sqrt{\frac{2}{n}} + \frac{1}{2n} \right). \tag{4.2}$$

Since the quantity on the right goes to 0 as $n \to \infty$, we have indeed shown that $\lim_{n \to \infty} N(n)/n = \pi$, as Tables 4.1 and 4.2 indicate.

The details of the above estimate for $N(n)$ have been nicely presented by Honsberger [5].

4.2 Sums of Two Squares

Gauss and others who followed him investigated many other aspects of lattice points associated with circles. As Dickson's *History of the Theory of Numbers* [2] demonstrates clearly, a lengthy and fascinating treatise could be written on this subject. To keep our discussion manageable, however, we will look at just a few of the classic lattice point problems and will not attempt many proofs. We continue our consideration of lattice points in circles by looking at how lattice points lead us to integer solutions of certain equations and inequalities related to circles and disks.

First, we give some definitions. Suppose that we can represent the nonnegative integer n as a *sum of two squares*; in other words, suppose that n can be written as $n = p^2 + q^2$, where the integers p and q can be positive, negative, or zero. The *number of representations of n as $p^2 + q^2$* is denoted by

$$R(n) = R(n = p^2 + q^2).$$

Each such representation $n = p^2 + q^2$ is considered *distinct* (that is, each is counted separately) even if the (p, q) pairs differ only in their sign or their order. We make one exception to this rule: $0 = 0^2 + 0^2$ is counted as a single representation. Here are some examples of distinct representations and the resulting $R(n)$ in each case:

$$R(0) = 1, \quad \text{since } 0 = 0^2 + 0^2,$$
$$R(4) = 4, \quad \text{since } 4 = (\pm 2)^2 + 0^2 = 0^2 + (\pm 2)^2,$$
$$R(8) = 4, \quad \text{since } 8 = (\pm 2)^2 + (\pm 2)^2,$$
$$R(10) = 8, \quad \text{since } 10 = (\pm 1)^2 + (\pm 3)^2 = (\pm 3)^2 + (\pm 1)^2.$$

Note that we have taken the ordering into account, as well as the distinct sign pairings $++$, $+-$, $-+$, $--$.

Now let's look at a small table of additional values for $R(n)$. As Table 4.3 shows, the values for $R(n)$ become very irregular as n increases. The values of $R(n)$ can be arbitrarily large, yet $R(n) = 0$ for infinitely many

Table 4.3. The Number of Integer Representations $R(n) = R\left(n = p^2 + q^2\right)$

n	Lattice point solutions (p, q)	$R(n)$	$T(n)$	$T(n)/n$
0	$(0, 0)$	1	1	
1	$(1, 0), (-1, 0), (0, 1), (0, -1)$	4	5	5.00
2	$(1, 1), (1, -1), (-1, 1), (-1, -1)$	4	9	4.50
$3 = 4 \cdot 0 + 3$		0	9	3.00
$4 = 2^2$	$(2, 0), (-2, 0), (0, 2), (0, -2)$	4	13	3.25
$5 = 4 \cdot 1 + 1$	$(2, 1), (-2, 1), (2, -1), (-2, -1),$			
	$(1, 2), (-1, 2), (1, -2), (-1, -2)$	8	21	4.20
$6 = 2 \cdot (4 \cdot 0 + 3)$		0	21	3.50
$7 = 4 \cdot 1 + 3$		0	21	3.00
$8 = 2^3$	$(2, 2), (-2, 2), (2, -2), (-2, -2)$	4	25	3.13
$9 = 3^2 = (4 \cdot 0 + 3)^2$	$(3, 0), (-3, 0), (0, 3), (0, -3)$	4	29	3.22
$10 = 2 \cdot (4 \cdot 1 + 1)$	$(\pm 3, \pm 1), (\pm 1, \pm 3)$	8	37	3.70
$11 = 4 \cdot 2 + 3$		0	37	3.36
$12 - 2^2 \cdot (4 \cdot 0 + 3)$		0	37	3.08
$13 = 4 \cdot 3 + 1$	$(\pm 2, \pm 3), (\pm 3, \pm 2)$	8	45	3.46
$14 = 2 \cdot (4 \cdot 1 + 3)$		0	45	3.21
$15 = 5 \cdot (4 \cdot 0 + 3)$		0	45	3.00
$16 = 2^4$	$(\pm 4, 0), (0, \pm 4)$	4	49	3.06
$17 = 4 \cdot 4 + 1$	$(\pm 4, \pm 1), (\pm 1, \pm 4)$	8	57	3.35
$18 = 2 \cdot (4 \cdot 0 + 3)^2$	$(\pm 3, \pm 3)$	4	61	3.39
$19 = 4 \cdot 4 + 3$		0	61	3.21
$20 = 2^2 \cdot (4 \cdot 1 + 1)$	$(\pm 4, \pm 2), (\pm 2, \pm 4)$	8	69	3.45
$21 = (4 \cdot 0 + 3) \cdot$		0	69	3.29
$(4 \cdot 1 + 3)$				
$22 = 2 \cdot (2 \cdot 4 + 3)$		0	69	3.13
$23 = 4 \cdot 5 + 3$		0	69	3.00
$24 = 2^3 \cdot (4 \cdot 0 + 3)$		0	69	2.88
$25 = (4 \cdot 1 + 1)^2$	$(\pm 5, 0), (0, \pm 5), (\pm 3, \pm 4), (\pm 4, \pm 3)$	12	81	3.24

values of n. That is, many integers have a large number of representations as a sum of two squares, while infinitely many integers have no such representation at all. We shall prove this latter statement the next section. In cases of such irregular behavior as $n \to \infty$, number theorists often seek an estimate for certain "average" values of these numerical functions.

We find such an average value in the usual way, that is, by summing the first n values and dividing by n. Explicitly, set

$$T(n) = R(0) + R(1) + R(2) + \cdots + R(n)$$

and consider the associated average:

$$\frac{T(n)}{n} = \frac{R(0) + R(1) + R(2) + \cdots + R(n)}{n}.$$

In a moment we will see how to use this average value. First, let's get back to our geometric point of view.

Observe that the circle $C(\sqrt{n})$ bounds the disk $D(\sqrt{n})$, defined as

$$D(\sqrt{n}) : p^2 + q^2 \leq n.$$

Each lattice point (p, q) in $D(\sqrt{n})$ gives an integer solution of this inequality.

How many integer solutions are there? Consulting Table 4.3, we can count the number of solutions of each of the equations

$$p^2 + q^2 = 0, \quad p^2 + q^2 = 1, \quad p^2 + q^2 = 2, \ldots, \quad p^2 + q^2 = n.$$

They are, respectively,

$$R(0) = 1, \quad R(1) = 4, \quad R(2) = 4, \ldots, \quad R(n).$$

Thus the sum $T(n) = R(0) + R(1) + \cdots + R(n)$ is exactly $N(n)$, the number of lattice points (p, q) inside or on the circle $C(\sqrt{n})$. For example, Table 4.3 shows that

$$T(4) = R(0) + R(1) + \cdots + R(4) = 1 + 4 + 4 + 0 + 4 = 13.$$

As Figure 4.2 illustrates, this total corresponds exactly to the number of lattice points on the circles $C(0)$, $C(\sqrt{1})$, $C(\sqrt{2})$, $C(\sqrt{3})$, and $C(\sqrt{4})$.

Furthermore, comparing Table 4.3 to Table 4.1, we see that each

$$T(n) = R(0) + R(1) + \cdots + R(n),$$

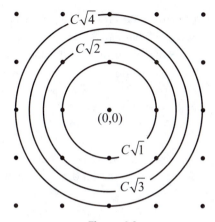

Figure 4.2

Lattice points in circles of radii $\sqrt{n}$, for $n = 1, 2, 3, 4$.

for $n = 1, 4, 9, 16$, and 25, corresponds exactly to each $N(n)$ for radii $\sqrt{n} = 1, 2, 3, 4$, and 5. Finally, since $R(n) = N(n)$, the average

$$\frac{R(n)}{n} = \frac{N(n)}{n} \to \pi \quad \text{as } n \to \infty,$$

from inequality (4.2) and our conclusion in Section 4.1. This limiting behavior does not show up clearly in Table 4.3, however, as n reaches only 25 there.

4.3 Numbers Representable as a Sum of Two Squares

The evidence from Table 4.3 shows that not all nonnegative integers n can be expressed in the form $n = p^2 + q^2$; for instance, $R(3) = 0$. Notice also that the circle $C(\sqrt{3})$ in Figure 4.2 has no lattice points on it. These related observations raise two important questions:

1. *Is there a way of characterizing nonnegative integers that can or cannot be represented as the sum of two squares?*

2. *If n can be represented as the sum of two squares, is there a formula that gives the total number of such representations?*

Both of these questions are very old. The *Arithmetic* of Diophantus, written around A.D. 250, addresses them, although the meaning of the statements there is unclear. The answer to question (1) was first stated by the Dutch mathematician Albert Girard (1595–1632) in 1625. A little later, Pierre de Fermat (1601–1665) also gave an answer, but without proof. (Of course, Fermat's reputation is such that no one would doubt the validity of his reasoning.) The first known proofs were published by Leonhard Euler (1707–1783) in 1749.

It is easy to show that integers of certain forms cannot be represented as a sum of two squares. Besides the example $R(3) = 0$ just cited, Table 4.3 shows that $R(n) = 0$ for $n = 7, 11, 15, 19, 23$. These particular integers have in common the property that all can be written in the form $4k + 3$. This property is crucial, in fact, as the following theorem demonstrates.

Theorem 4.1. *No integer of the form* $4k + 3$, *where* $k = 0, 1, 2, \ldots$, *can be represented as a sum* $p^2 + q^2$ *with integers* p, q.

Proof. If p is even (say, $p = 2h$), then $p^2 = 4h^2$ is divisible by 4. But if p is odd (say, $p = 2h + 1$), then division by 4 leaves a remainder of 1:

$$p^2 = (2h + 1)^2 = 4h^2 + 4h + 1 = 4(h^2 + h) + 1.$$

It follows that dividing the sum $p^2 + q^2$ by 4 leaves:

1. no remainder (if both p and q are even);

2. a remainder of 1 (if one of them is even, the other odd); or

3. a remainder of 2 (if both are odd).

In no case does $p^2 + q^2$ have 3 as the remainder upon division by 4, so an integer of the form $4k + 3$ cannot be represented as a sum of two squares.

Theorem 4.1 shows why every fourth entry for $R(n)$ in Table 4.3 must be 0, no matter how far we continue. But why are there so many other 0 entries? We shall find out why by examining more closely some theorems about which numbers can or cannot be represented as sums of squares.

We begin with the Unique Factorization Theorem; for a proof, see Niven [7]. This theorem states that every positive integer $n > 1$ can be represented in one and only one way in the so-called *standard form*:

$$n = 2^\beta p_1^{\alpha_1} p_2^{\alpha_2} \cdots p_s^{\alpha_s},$$

where $p_1, p_2, \ldots, p_s$ are distinct odd primes and where none of the exponents $\alpha_1, \alpha_2, \ldots, \alpha_s$ is zero. If n is odd, then $\beta = 0$; otherwise β is the highest power of 2 that divides n. With n so expressed, we can state the following important theorem.

Theorem 4.2. *A positive integer n is the sum of two squares if and only if all its prime factors of the form $4k + 3$ have even exponents when n is expressed in standard form.*

For example, the positive integers $3, 6, 7, 11, 12, 14, 15, 19, 21, 22, 23$, and 24 cannot be sums of two squares, but 9 and 18 can be so expressed.

We will not prove Theorem 4.2 here, for it is not easy; the interested reader may consult Hardy and Wright [**4**]. We can, however, pluck out one important ingredient from the proof of Theorem 4.2 to give a taste of the recipe. This ingredient is the identity

$$(a^2 + b^2)(c^2 + d^2) = (ac + bd)^2 + (ad - bc)^2, \qquad (4.3)$$

known to Leonardo Fibonacci (c. 1170–c. 1250), who gave it in his book *Liber Abaci* in 1202. This identity shows that when each of two integers is a sum of two squares, their product is itself a sum of two squares.

The following examples illustrate how this identity is used in the proof of Theorem 4.2. Referring to Table 4.3, we see that

$$4 = 2^2 = (\pm 2)^2 + 0^2 = 0^2 + (\pm 2)^2$$

and

$$9 = 3^2 = (\pm 3)^2 + 0^2 = 0^2 + (\pm 3)^2.$$

Letting $a = 2$, $b = 0$, $c = 3$, and $d = 0$ and applying the identity, we get

$$4 \cdot 9 = 36 = (2 \cdot 3 + 0 \cdot 0)^2 + (2 \cdot 0 - 0 \cdot 3)^2 = 6^2 + 0^2.$$

Thus, we have constructed a representation of 36 as a sum of two squares. All other representations are obtained in the same way, by combining the various orderings and signs in the representations of 4 and 9. Similarly,

$$18 = 2 \cdot 3^2 = (1^2 + 1^2)(3^2 + 0^2) = 3^2 + 3^2$$

gives a representation of 18 as a sum of two squares.

In fact, suppose we had shown that primes of form $4k + 1$ can be written as a sum of two squares. Then it follows from (4.3) that a product of such primes can also be so represented. A sum of two squares multiplied by a perfect square can again be represented as a sum of two squares. Since a product of prime factors with even exponents is a perfect square, this shows that the condition stated in Theorem 4.2 is sufficient.

Problem Set for Section 4.3

1. Show that Theorem 4.2 implies Theorem 4.1.
2. The absolute value of the product of two complex numbers is the product of their absolute values. Using this fact, deduce the identity (4.3).

4.4 Representations of Prime Numbers as Sums of Two Squares

As we have learned from Theorem 4.2, certain primes can be written as a sum of two squares. A famous, deep theorem addresses the number of ways in which this can be done.

Theorem 4.3. *Every prime of the form $l = 4k + 1$ can be written as the sum $p^2 + q^2 = l$, and only one pair of integers p and q (where $0 < p < q$) supplies this representation.*

The condition $0 < p < q$ simply means that we ignore the order and the signs. Thus, for the representations in Table 4.3, such a pair p, q would account for the entries $R(4k+1) = 8$ whenever $4k+1$ is a prime, counting each arrangement

$$(p, q) \ (-p, q) \ (p, -q) \ (-p, -q) \ (q, p) \ (-q, p) \ (q, -p) \ (-q, -p)$$

only once.

Theorem 4.3 has a long history [1, 2, 8]. Fermat noted it in the margin of his well-worn copy of the works of Diophantus, and he also stated it in a letter dated December 25, 1640, bearing his mathematical Christmas greetings to Friar Marin Mersenne (1588–1648), a great scholar of primes. As we have said, the first published proof was due to Euler. Another proof, employing simple theorems of congruences, is due to the Norwegian mathematician Alex Thue (1863–1922), who made important contributions to the modern theory of Diophantine equations.

Theorem 4.3 says that the representation can be written in only one way. Where does this uniqueness statement come from? It follows from Euler's observation that if an odd integer can be represented as the sum of two squares in two different ways, then it can be factored, and so it cannot be a prime. Here is Euler's proof.

Proof by Euler. Suppose that we represent the prime $l = 4k + 1$ in two different ways: as the sum $l = a^2 + b^2$ and also as $l = c^2 + d^2$, with a, c odd, b, d even, and $a, c, b, d > 0$. The two representations will lead to a factorization of l, contradicting the assumption that l is a prime.

If we have $l = a^2 + b^2 = c^2 + d^2$, then we can factor

$$(a - c)(a + c) = (d + b)(d - b). \tag{4.4}$$

Since $a \neq c$ and $b \neq d$, no factor in equation (4.4) is zero. Letting $k = \text{g.c.d.}(a - c, b - d)$, we write

$$a - c = ks \quad \text{and} \quad b - d = kt, \quad \text{where g.c.d.}(s, t) = 1.$$

Since both $a - c$ and $b - d$ are even, 2 certainly divides both of them. Hence, 2 must be a factor of k; so k must be even.

Substituting into (4.4), we get $ks(a + c) = kt(b + d)$, or

$$s(a + c) = t(b + d). \tag{4.5}$$

Because g.c.d.$(s, t) = 1$, we must have that $t|(a + c)$ and $s|(b + d)$. Then, from (4.5), we see that

$$\frac{a + c}{t} = \frac{b + d}{s} = n.$$

Therefore,

$$\begin{aligned} a + c &= nt, \\ b + d &= ns. \end{aligned} \tag{4.6}$$

Clearly, n is a common divisor of $(a + c)$ and $(b + d)$, so $n|(a + c, b + d)$. We claim that $n = (a + c, b + d)$. For suppose that

$$nr = (a + c, b + d).$$

Then there exist integers j_1 and j_2 such that

$$nrj_1 = a + c \quad \text{and} \quad nrj_2 = b + d.$$

Substituting these equivalences into (4.6), we see that

$$nrj_1 = nt \quad \text{and} \quad nrj_2 = ns,$$

or that

$$rj_1 = t \quad \text{and} \quad rj_2 = s.$$

But this means that $r \mid$ g.c.d.$(s, t) = 1$, and so $r = 1$. Hence $n = $ g.c.d. $(a + c, b + d)$; and since $a + c$ and $b + d$ are both even, n is also even.

Now, using the fact that k and n are even, we can factor l as a product of integers in the form

$$l = \left[\left(\frac{k}{2} \right)^2 + \left(\frac{n}{2} \right)^2 \right] (s^2 + t^2),$$

where neither factor is 1 since k, n, s, t are all nonzero.

To check the factorization, we need only multiply out the right-hand side, obtaining

$$\frac{1}{4} \left[(kt)^2 + (ks)^2 + (tn)^2 + (ts)^2 \right]$$

$$= \frac{1}{4} \left[(d - b)^2 + (a - c)^2 + (a + c)^2 + (d + b)^2 \right]$$

$$= \frac{1}{4} \left[2 \left(a^2 + b^2 \right) + 2 \left(c^2 + d^2 \right) \right]$$

$$= \frac{1}{4} (2l + 2l)$$

$$= l.$$

This contradicts our hypothesis that l is prime, so Theorem 4.3 is proved.

4.5   A Formula for *R(n)*

At the beginning of Section 4.3 we posed a question:

If a number n can be represented as the sum of two squares, is there a formula that gives the total number of such representations?

In short, does a formula for $R(n) = R(n = p^2 + q^2)$ exist?

This question was answered by Adrien Marie Legendre (1752–1833) in the following theorem. Our wording uses modern terminology, including the notation $u \equiv v \pmod{m}$ to mean that the difference $u - v$ is divisible by m.

Theorem 4.4. *Suppose that $n \geq 1$ has A divisors $\alpha_1, \ldots, \alpha_A$, where $\alpha_i \equiv 1 \pmod 4$, and B divisors $\beta_1, \ldots, \beta_B$, where $\beta_j \equiv -1 \pmod 4$. Then $R(n) = 4(A - B)$.*

Table 4.4. Examples Illustrating Theorem 4.4

n	$R(n)$	Divisors of n	A^a	B^b	$4(A - B)$
2	4	$1, 2$	1	0	$4(1) = 4$
5	8	$1, 5$	2	0	$4(2) = 8$
7	0	$1, 7$	1	1	$4(0) = 0$
65	16	$1, 5, 13, 65$	4	0	$4(4) = 16$
200	12	$1, 2, 4, 5, 8, 10, 20,$			
		$25, 40, 50, 100, 200$	3	0	$4(3) = 12$

^{a} Number of divisors $\equiv 1 \pmod 4$.
^{b} Number of divisors $\equiv -1 \pmod 4$.

Note that we are talking about all divisors, not merely prime divisors. For example, examine Table 4.4.

Readers interested in a proof of Theorem 4.4 are referred to the problem set at the end of this section, which contains a sketch of Jacobi's 1834 proof.

Now, expanding on the previous question, we ask:

Does a formula exist for evaluating the sum $T(n) = R(0) + R(1) + \cdots + R(n)$?

The answer to this second question comes from a manuscript by Gauss [3] published posthumously. He proved that the exact value of this sum is given by the formula

$$T(n) = 1 + 4 \left\{ \left[\frac{n}{1} \right] - \left[\frac{n}{3} \right] + \left[\frac{n}{5} \right] - \left[\frac{n}{7} \right] + \cdots \right\}, \qquad (4.7)$$

where $[t]$, as in Chapter 2, denotes the largest integer less than or equal to t. In particular, notice that $[n/(2k + 1)] = 0$ if $2k + 1 > n$.

Clearly, for large values of n formula (4.7) is not practical. Here is a more efficient formula:

$$T(n) = 1 + 4 \sum_{k=0}^{[\sqrt{n}]} \left[\sqrt{n - k^2} \right]. \qquad (4.8)$$

For example, using (4.8) we can find the sum

$$T(100) = 1 + 4\left\{ \left[\sqrt{100}\right] + \left[\sqrt{99}\right] + \left[\sqrt{96}\right] + \left[\sqrt{91}\right] + \left[\sqrt{84}\right] + \right.$$
$$\left. \left[\sqrt{75}\right] + \left[\sqrt{64}\right] + \left[\sqrt{51}\right] + \left[\sqrt{36}\right] + \left[\sqrt{19}\right] \right\}$$
$$= 1 + 4(10 + 9 + 9 + 9 + 9 + 8 + 8 + 7 + 6 + 4) = 317.$$

Sierpenski [9] gives a proof of formula (4.8) and shows its connection with Gauss's formula (4.7). The equivalence of the two formulas is known as *Liouville's identity*, after Joseph Liouville (1809–1882).

Problem Set for Section 4.5

1. Is it true that if $n = 4k + 6$, then $R(n) = 0$? If not, give a counterexample.
2. Prove that if $n = 12k + 9$, and 3 does not divide k, then $R(n) = 0$.
3. Evaluate $R(1225) = R(5^2 \cdot 7^2)$.
4. Evaluate $T(1225)$.
5. To calculate solutions p, q of the equation $n = p^2 + q^2$ (or of $n - p^2 = q^2$), it suffices to substitute for p the values $p = 0, 1, 2, 3, \ldots$, whose absolute values are less than or equal to $\sqrt{n}$, and then to see whether the difference $n - p^2$ is (or is not) a square. Show that for any fixed n the difference of consecutive terms of the sequence $n - 0^2$, $n - 1^2$, $n - 2^2, \ldots$ is the sequence of consecutive odd integers $1, 3, 5, \ldots$.
6. Use the ideas of Problem 5 to calculate the values of $R(5)$, $R(10)$, and $R(25)$.
7. *(Optional)* In 1834, Karl Gustav Jacob Jacobi (1804–1851) proved the formula $R(n) = 4(A - B)$ in Theorem 4.4. His proof was not elementary, being based on an identity obtained from his profound study of elliptic functions. We will sketch it lightly here.

   Proof by Jacobi. Jacobi proved the identity

   $$\left(1 + 2x + 2x^4 + 2x^9 + 2x^{16} + \cdots\right)^2$$
   $$= 1 + 4\left(\frac{x}{1 - x} - \frac{x^3}{1 - x^3} + \frac{x^5}{1 - x^5} - \cdots\right).$$

   By the rules of elementary algebra, squaring the left side of this identity gives terms that can be arranged in the form of a power series,

   $$1 + a_1 x + a_2 x^2 + a_3 x^3 + \cdots + a_n x^n + \cdots;$$

so then $a_n = R(n)$. Similarly, if the right-hand side is expanded and expressed in the form

$$1 + b_1 x + b_2 x^2 + b_3 x^3 + \cdots,$$

then $b_n = 4(A - B)$. Finally, equating the coefficients of like powers of x in both of these expansions yields $R(n) = 4(A - B)$.

Try Jacobi's proof of $R(n) = 4(A - B)$ using $a_1, a_2, \ldots, a_5$ and $b_1, b_2, \ldots, b_5$. Note that

$$\frac{x}{1 - x} = x + x^2 + x^3 + \cdots,$$

$$\frac{x^2}{1 - x^2} = x^2 + x^4 + x^6 + \cdots,$$

and so on.

References

1. L. E. Dickson, "Methods of Factoring," Chapter 14 in *History of the Theory of Numbers, Vol. I: Divisibility and Primality* (Washington, D.C.: Carnegie Institute, 1919), 360.
2. _____, "Sum of Two Squares," Chapter 6 in *History of the Theory of Numbers, Vol. II: Diophantine Analysis* (Washington, D.C.: Carnegie Institute, 1920), 225.
3. C. F. Gauss, *Werke* (Göttingen: Gesellschaft der Wissenschaften, 1863–1933).
4. G. H. Hardy and E. M. Wright, Chapter 10, Theorem 366, in *An Introduction to the Theory of Numbers*, 5th ed. (Oxford: Oxford University Press, 1983).
5. Ross Honsberger, "Writing a Number as a Sum of Two Squares," Essay 8 in *Ingenuity in Mathematics*, New Mathematical Library Series, Vol. 23 (New York: Random House, 1970), 61–66.
6. H. L. Mitchell III, *Numerical Experiments on the Number of Lattice Points in the Circle* (Stanford, CA: Stanford University, Applied Mathematics and Statistics Labs, 1961).
7. Ivan Niven, Appendix B in *Numbers: Rational and Irrational*, New Mathematical Library Series, Vol. 1 (New York and Toronto: Random House, 1961).
8. Oystein Ore, *Number Theory and Its History* (New York: McGraw-Hill, 1948; reprinted with supplement, New York: Dover, 1988).
9. W. Sierpinski, *Elementary Theory of Numbers*, 2nd ed., Andrzej Schinzel, ed., North-Holland Mathematical Library, Vol. 31 (Amsterdam and New York: North-Holland; Warsaw: Polish Scientific Publishers, 1988).

Part II

An Introduction to the Geometry of Numbers

5

Minkowski's
Fundamental Theorem

5.1 Minkowski's Geometric Approach

As we said in Section 1.1, the *geometry of numbers* is an important branch of number theory that originated in the work of Hermann Minkowski. The reader will find a biographical sketch of this great mathematician in Appendix III.

The geometry of numbers is connected with the problem of determining whether inequalities of various kinds are solvable in integers. Inequalities for which integer solutions are sought are called *Diophantine inequalities*. Earlier, using algebraic methods, Charles Hermite (1822–1901) had proved many general theorems on the solutions of Diophantine inequalities, the most important of which he communicated to Karl Jacobi in letters written about 1845. Like Hermite, Minkowski was interested in such problems, but he approached them quite differently from anyone who had come before him.

Minkowski's was a geometrical point of view. He identified simple geometric conditions under which certain regions in the plane contain lattice points. He also generalized his results to n-dimensional space, thus achieving new and simpler proofs of Hermite's algebraic results. When, around 1890, Minkowski wrote to Hermite to inform him of these accomplishments, Hermite expressed lively interest in Minkowski's discoveries.

It was Minkowski who christened this new field of study the *geometry of numbers*. His exposition of this subject appeared in two books, *Geometrie der Zahlen* (1896) [6] and *Diophantische Approximationen* (1907) [7], the latter of which is easier to read. For a modern, deep, and powerful account, see Cassels [1].

Here is an example of the type of problem that fascinated these great mathematicians:

For any given real number α, are there integers m and n, $m \neq 0$, such that $|\alpha - (n/m)| \leq 1/2m$?

One way to answer this question is to consider an arbitrary integer $m > 1$ and take n to be the integer closest to αm. We know that

$$|\alpha m - n| \leq \frac{1}{2}$$

so

$$\left| \alpha - \frac{n}{m} \right| \leq \frac{1}{2m}.$$

Thus, infinitely many pairs of integers m, n in fact satisfy this inequality. The larger we take m, the better will be our approximation. Moreover, if α is irrational, then the strict inequalities hold. Why? Because then αm is irrational, too, so it cannot lie exactly half a unit from an integer.

Taking Minkowski's viewpoint, we can state this fact geometrically:

The strip bounded by the straight lines $\alpha x - y = \frac{1}{2}$ and $\alpha x - y = -\frac{1}{2}$ contains infinitely many lattice points.

Although this strip may be extremely narrow, the result still holds—and that fact should not surprise us. As we learned in Section 1.6, any rectangle symmetric with respect to the origin will contain lattice points besides $(0,0)$ as soon as its area exceeds 4.

Suppose we want a better rational approximation of a given real number α. For which (if any) special integral values m is αm particularly close to an integer n? We shall show in Chapter 6 that, indeed, for infinitely many integers m, n, $m \neq 0$, the difference $\alpha - (n/m)$ satisfies the inequality

$$\left| \alpha - \frac{n}{m} \right| < \frac{1}{2m^2}. \tag{5.1}$$

See Problem 2 for a geometric interpretation.

Problem Set for Section 5.1

1. Consider the strip bounded by the lines $y - \alpha x = \frac{1}{2}$ and $y - \alpha x = -\frac{1}{2}$.

   a. Show that it is symmetric with respect to the origin.

   b. Express its width as a function of α.

c. Find the constant k, as a function of α, such that the lines $x+\alpha y = k$ and $x + \alpha y = -k$ cut from the strip the smallest rectangle that contains a lattice point other than $(0,0)$ in its interior or on its boundary.

d. Answer parts (b) and (c) in the case $\alpha = \sqrt{3}$.

2. Multiplying inequality (5.1) by $2m^2$ gives us the equivalent inequalities

$$-1 \leq 2\alpha m^2 - 2mn \leq 1. \tag{5.1$'$}$$

For given α, infinitely many integers m, n exist such that (5.1) is satisfied. This assertion is equivalent to saying that the region S described by the condition (5.1$'$) contains infinitely many lattice points.

a. Show that S is symmetric with respect to the origin.

b. Verify that S is the region lying between two conjugate hyperbolas with asymptotes $x = 0$ and $y - \alpha x = 0$.

5.2 Minkowski M-Sets

Minkowski introduced a key figure in the plane that we shall call an M-set. The precise shape of an M-set will depend on the particular problem to be solved, but the following two properties are both required.

Property 1. An M-set is convex.

We call a set of points *convex* if it contains every point on the line segment connecting any two of its points; see Figure 5.1.

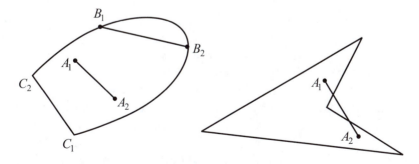

Figure 5.1
Left: Convex sets. *Right:* Nonconvex sets.

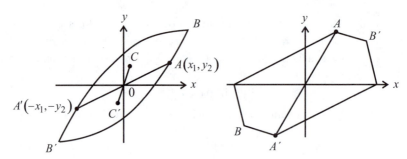

Figure 5.2
Two M-sets centered at $(0,0)$.

Sometimes Minkowski used a second definition of convexity, namely: An M-set is convex if it is possible to draw a line ℓ through every point P on the boundary of M, such that the whole of M lies on one side of ℓ.

Property 2. An M-set is symmetric with respect to a point O.

That is, if P is a point of the set, then the point P' on the line through P and O, and such that $|\overline{P'O}| = |\overline{OP}|$, also belongs to the set.

The point O is called the set's *center of symmetry*. For convenience, we shall often make the origin $(0,0)$ the center of an M-set, as in Figure 5.2.

A consequence of property (2) is that whenever such a set contains a point with coordinates (a,b), it also contains the point with coordinates $(-a,-b)$. Clearly, the origin bisects the line segment $\overline{AA'}$ connecting a point with its symmetric image.

It is easy to *expand* or *contract* an M-set with center $(0,0)$. We simply map each (x,y) of the set into the point (tx,ty) , where t is a real number. This resizing yields a similar M-set that is, likewise, symmetric with respect to the origin; see Figure 5.3.

While we have considered only planar models so far, it should be clear that the definition of M-sets can be generalized to spaces of three and higher dimensions. For example, all cubes, spheres, and ellipsoids centered at the origin are M-sets in 3-space. Until we have mastered Minkowski's concepts in the plane, however, we will defer the more difficult considerations of higher dimensions. For an excellent reference on convex figures and polyhedra, see Lyusternik [5].

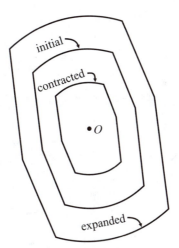

Figure 5.3
Dilation retains symmetry.

Problem Set for Section 5.2

1. A convex set Q contains three non-collinear points A, B, C. Prove that Q then contains all of the triangle ABC.

2. Minkowski proved that if a convex polygon Q can be decomposed into a finite number of centrally symmetric polygons, then Q has central symmetry.

   a. Draw figures to illustrate this theorem.

   b. What happens when we leave out the word "convex"? Illustrate.

3. Prove that the intersection of two convex point sets in the plane is convex.

5.3 Minkowski's Fundamental Theorem

We now state and prove the fundamental theorem in the geometry of numbers.

Theorem 5.1 (Minkowski's Fundamental Theorem). *Let C be a two-dimensional M-set with center at the origin O and area greater than or equal to 4. Then C contains, either in its interior or on its boundary, lattice points of Λ other than O.*

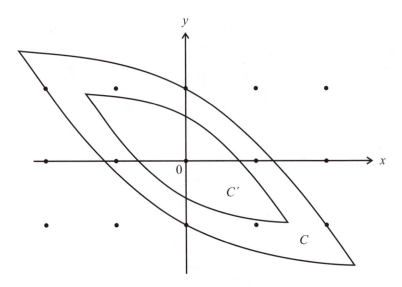

Figure 5.4

The M-set C contracted to C'.

Proof. We consider the M-set C with center at $(0,0)$ and area $A > 4$. We contract C by mapping every point (x, y) of C into the point $(x/2, y/2)$; see Figure 5.4. This contraction yields an M-set C' similar to C and such that lengths of segments in C' are just half of the lengths of corresponding segments in C. Hence the Area A' of C' satisfies

$$A' = \left(\frac{1}{2}\right)^2 A = \frac{1}{4}A,$$

and since $A > 4$, $A' > 1$.

Next we place a replica of C' on every lattice point of Λ. In other words, we *translate* the M-set C' from the origin to every point (p, q) with integer coordinates. Thus, if (x', y') is a point of C', then $(x' + p, y' + q)$ is the corresponding point of the translate of C' centered at (p, q).

Examine these translates in Figure 5.5.

Do they appear to overlap? Do some points in the plane seem to belong to more than one translate?

We shall show that these translates indeed overlap and that, as a consequence, our original M-set C must have contained a lattice point other than the origin.

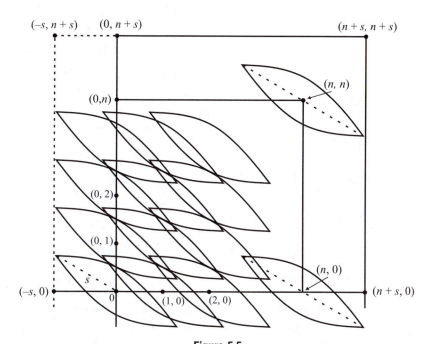

Figure 5.5
The translates of C' overlap.

We begin by considering the square in Figure 5.5 with vertices $(0,0)$, $(n,0)$, (n,n), $(0,n)$, where n is some positive integer. The square has $(n+1)^2$ lattice points in its interior and on its boundary. The sum of the areas of the $(n+1)^2$ translates of C' centered on these lattice points is $(n+1)^2 A'$.

Let s be the maximum distance of any point of C' from the center $(0,0)$. Then all these $(n+1)^2$ M-sets are contained in a square of side $n+2s$ and of area $(n+2s)^2$. Again, see Figure 5.5. As we shall show, the sum of the areas of the $(n+1)^2$ translates of C' exceeds the area of the square containing them; that is, we shall show that

$$(n+1)^2 A' > (n+2s)^2. \tag{5.2}$$

From this, we will conclude that the translates must overlap.

To prove inequality (5.2), we subtract $(n+2s)^2$ from both sides and prove the equivalent inequality

$$(n+1)^2 A' - (n+2s)^2 = (A'-1)n^2 + 2n(A'-2s) + A' - 4s^2 > 0. \tag{5.3}$$

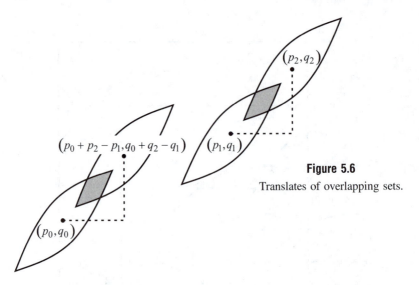

Figure 5.6

Translates of overlapping sets.

Since $A' > 1$, the leading coefficient of the quadratic expression in n is positive, and the entire expression (5.3) is therefore positive for sufficiently large n. If we choose n so large that (5.3) is positive, then $(n + 1)^2 A' > (n + 2s)^2$ and the translates of C' overlap in the square.

Do you see why every translate of C' has points in common with some other translate? Consider the two overlapping sets centered at (p_1, q_1) and (p_2, q_2) as in Figure 5.6. A third translate centered at any lattice point (p_0, q_0) must then have points in common with a fourth centered at $(p_0 + p_2 - p_1, q_0 + q_2 - q_1)$. In particular, C' itself, centered at $(0, 0)$, has points in common with the translate C'' centered at (p, q), where $p = p_2 - p_1, q = q_2 - q_1$; see Figure 5.7. Thus every point (x'', y'') of C''

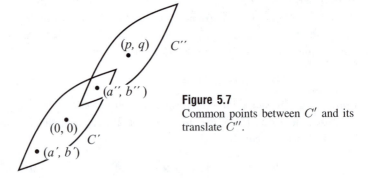

Figure 5.7

Common points between C' and its translate C''.

may be written

$$x'' = x' + p, \qquad y'' = y' + q,$$

where (x', y') is the corresponding point of C'. Hence, if two translates overlap, then every translate must overlap at least one other.

Now, consider any point (a'', b'') that is in C' and also in C'''. A point (a', b') must exist in C' such that

$$a'' = a' + p, \qquad b'' = b' + q. \tag{5.4}$$

Because C' is symmetric, C' also contains the point $(-a', -b')$. And because C' is convex, it contains the midpoint of the segment connecting any two of its points. The midpoint connecting points (a'', b'') and $(-a', -b')$ has coordinates

$$\left(\frac{a'' - a'}{2}, \frac{b'' - b'}{2} \right),$$

which by (5.4) is the point $(p/2, q/2)$. Therefore C' contains the point $(p/2, q/2)$, where (p, q) are the integers associated with the center of C''. It follows that the original M-set C contains the lattice point (p, q).

Is our proof complete? Not quite, because we assumed that the area of C was greater than 4, whereas Theorem 5.1 asserts that C contains a lattice point in its interior or on its boundary even if $A = 4$.

So, letting $A = 4$, suppose that C has no lattice points other than $(0, 0)$ either inside or on its boundary. Then every lattice point is farther than some positive distance δ from any point of C. We now expand C so slightly that all the points of the resulting enlarged M-set C^* are still at least $\delta/2$ away from the nearest lattice point. Now C^* has area greater than 4, but it still contains no lattice points except $(0, 0)$. However, this contradicts the results just obtained. Thus C must have had a lattice point other than the origin either in its interior or on its boundary. This completes the proof of Minkowski's Fundamental Theorem.

What we have presented is essentially one of Minkowski's own original proofs. Other proofs exist; see Koksma's list [4] and, for one of the most interesting proofs, by Hajós [2], see Hardy and Wright [3]. The proof by Mordell [8] is also recommended.

Problem Set for Section 5.3

1. Consider an M-set C with area 6, where the maximum distance from $(0,0)$ to a point of C is 5. Find the smallest positive integer n such that $(n+1)^2 A' > (n+2s)^2$, where A' and s are the quantities defined in the text.

2. Using straight lines, draw a nonconvex figure with symmetry about $(0,0)$, area $A > 4$, and no lattice points except $(0,0)$ inside it or on its boundary.

3. Here is another proof of Minkowski's Fundamental Theorem, due to Blichfeldt. It is like ours, except for our argument establishing a pair of points in C' whose coordinates differ by integers; see relations (5.4). Instead, we will consider the M-set C' of area $A' > 1$, which is cut into pieces by the lattice L as shown in Figure 5.8.

   We fit each piece into a unit square U in such a way that it occupies the same position in U as it occupied in the lattice square from which it came. Since $A' > 1$, while the area of $U = 1$, some pieces of C' overlap in U. Let D_1 and D_2 be two pieces of C' that overlap in U,

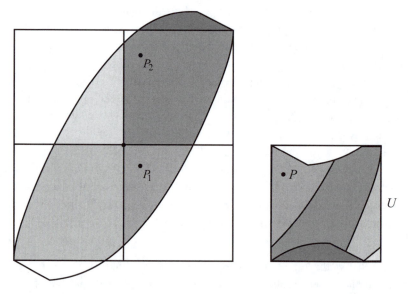

Figure 5.8
Lattice square and unit square.

and let P_1, P_2 be points of D_1 and D_2 that coincide when D_1 and D_2 occupy U.

   a. Complete this proof by showing that the coordinates of P_1 differ from those of P_2 by integers.

   b. Use the symmetry and convexity of C' as done in the text to establish the existence of a lattice point in C other then $(0,0)$.

4. Study assignment: Study the proof of Minkowski's Fundamental Theorem given by Mordell.

5.4 (Optional) Minkowski's Theorem in *n* Dimensions

Minkowski did not limit his theorems to the plane, but generalized to many spaces of n-dimensions. Here is an example of one, which we state without proof.

Theorem 5.2 (Minkowski's General Theorem). *Any convex set (or body) in n-dimensional space that is symmetrical about the origin and has volume greater than 2^n contains a point with integral coordinates $(x_1, x_2, \ldots, x_n)$, other than $(0, 0, \ldots, 0)$.*

In this general form, which applies to any symmetrical convex set, the number 2^n cannot be replaced by a smaller number. To prove this, we have only to consider the cube in n-space,

$$|x_1| < 1, |x_2| < 1, \ldots, |x_n| < 1.$$

This cube is clearly a convex symmetrical set whose volume is exactly 2^n, and it contains no lattice point other than the origin.

    We can also state another, more general, geometrical interpretation. Suppose we apply any linear transformation to the variables $x_1, x_2, \ldots, x_n$:

$$y_1 = \alpha_{11}x_1 + \alpha_{12}x_2 + \cdots + \alpha_{1n}x_n,$$

$$y_2 = \alpha_{21}x_1 + \alpha_{22}x_2 + \cdots + \alpha_{2n}x_n,$$

$$\vdots$$

$$y_n = \alpha_{n1}x_1 + \alpha_{n2}x_2 + \cdots + \alpha_{nn}x_n.$$

Here we want the coefficients α_{ij} to be real, and we require the matrix of coefficients to have nonzero determinant Δ. This represents the *general affine transformation* mapping the x-space into the y-space. The system of points with integral coordinates in x-space has been transformed into a system of points in y-space of a more general kind, which we call a lattice.

Let's look more closely at this lattice in y-space. Let $A_1 = (\alpha_{11}, \alpha_{21}, \alpha_{31}, \ldots, \alpha_{n1})$ denote the point in y-space that corresponds to the point $(1, 0, 0, \ldots, 0)$ in x-space; let $A_2 = (\alpha_{12}, \alpha_{22}, \ldots, \alpha_{n2})$ denote the point in y-space corresponding to the point $(0, 1, 0, \ldots, 0)$ in x-space; and so on. The general point P of the lattice can then be given in vector notation as

$$\overrightarrow{OP} = m_1 \overrightarrow{OA_1} + m_2 \overrightarrow{OA_2} + \cdots + m_n \overrightarrow{OA_n},$$

where $m_1, m_2, \ldots, m_n$ take on all integral values. The vectors $\overrightarrow{OA_1}, \overrightarrow{OA_2}, \ldots, \overrightarrow{OA_n}$ form a structure that is called the *fundamental parallelepiped* of the lattice.

A set (or body) of volume V in x-space is transformed by our affine transformation into a set of volume $V' = V \cdot |\Delta|$ in y-space. The fundamental parallelopiped has volume $|\Delta|$, which is called the *determinant of the lattice*. We can now restate Minkowski's Fundamental Theorem in y-space as follows.

Theorem 5.3 (Minkowski's Fundamental Theorem in y-space). *Let L be any lattice of determinant Δ in n-dimensional space. Then any convex body symmetrical about the origin whose volume exceeds $2^n \Delta$ contains a point of L other than $(0, 0)$.*

References

1. J. W. S. Cassels, *Introduction to the Geometry of Numbers*, in Classics of Mathematics Series (1971; corrected reprint, Berlin: Springer-Verlag, 1997).
2. G. Hajós, "Ein neuer Beweis eines Satzes von Minkowski," *Acta Litt. Sci. (Szeged)* 6 (1934):224-5.
3. G. H. Hardy and E. M. Wright, notes for Chapter 3 in *An Introduction to the Theory of Numbers*, 5th ed. (Oxford: Oxford University Press, 1983), 37.
4. J. F. Koksma, *Diophantische Approximationen* (New York: Chelsea, 1936), 13.
5. L. A. Lyusternik, *Convex Figures and Polyhedra*; 1st ed. (1956) translated from Russian and adapted by Donald L. Barrett (Boston: D. C. Heath, 1966).
6. Hermann Minkowski, *Geometrie der Zahlen*, Bibliotheca Mathematic Teubneriana, Vol. 40 (Leipzig: Teubner, 1910; New York and London: Johnson Reprint Corp., 1988). First section of 240 pages appeared in 1896.
7. ――――, *Diophantische Approximationen: Eine Einfuhrung in die Zahlentheorie* (reprinted, New York: Chelsea, 1957).
8. L. J. Mordell, "On Some Arithmetical Results in the Geometry of Numbers," *Compositio Math.* 1 (1934):248–53.

6
Applications of Minkowski's Theorems

6.1 Approximating Real Numbers

For further information on Hermann Minkowski's discoveries in the geometry of numbers, the interested reader of German should go to his selected papers [**4**]. There one will find how Minkowski delved deeply into this subject, investigating questions and proving theorems in three dimensions and higher. This chapter will explore some of the ways in which his results help us establish the accuracy possible in approximating real numbers by rational numbers.

As our first application of Minkowski's Fundamental Theorem, we shall prove the following approximation. It is illustrated by the parallelogram shown in Figure 6.1.

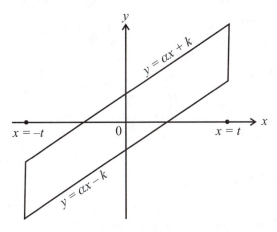

Figure 6.1
Parallelogram for proof of Theorem 6.1

Theorem 6.1.   *Given any real number α and an integer $t > 0$, however large, there exist integers p, q, not both zero, such that $|q - \alpha p| \leq 1/t$.*

Proof.   Take for our M-set the parallelogram in Figure 6.1 bounded by the four lines

$$y - \alpha x = k, \quad y - \alpha x = -k, \quad x = t, \quad x = -t.$$

This parallelogram has base $2t$, altitude $2k$, and, hence, area $A = 2t \cdot 2k = 4tk$. Thus, if we take $k = 1/t$, with t a positive integer, then the area $A = 4$. By Minkowski's Fundamental Theorem, there must be at least one lattice point (p, q) other than $(0, 0)$ within this parallelogram or on its boundary.

This statement implies two things about the values of p and q. First, observe that

$$-t \leq p \leq t, \quad \text{or} \quad |p| \leq t.$$

Second, it must be true that

$$\alpha p - k \leq q \leq \alpha p + k.$$

Because $k = 1/t$, we can rewrite the value of q as

$$\alpha p - \frac{1}{t} \leq q \leq \alpha p + \frac{1}{t}, \quad \text{where } t > 0.$$

Subtracting αp yields $-1/t \leq q - \alpha p \leq 1/t$. Thus, $|q - \alpha p| \leq 1/t$. This proves Theorem 6.1.

Remarks.   The theorem says that p, q cannot both be zero. If $p \neq 0$, then

$$\left| \frac{q}{p} - \alpha \right| \leq \frac{1}{|p|t}.$$

Here, when t is very large, $1/|p|t$ is very small. Thus this inequality gives excellent rational approximations q/p to any real number α.

6.2   Minkowski's First Theorem

We now apply Minkowski's Fundamental Theorem to another M-set to obtain lattice points where two linear forms can be simultaneously bounded. This result is usually called *Minkowski's First Theorem*, not to be confused with his *Fundamental Theorem*, 6.1.

Theorem 6.2 (Minkowski's First Theorem). *Consider two linear forms*

$$\xi = \alpha x + \beta y,$$
$$\eta = \gamma x + \delta y,$$

determinant $\Delta = \alpha\delta - \beta\gamma \neq 0,$

where $\alpha, \beta, \gamma, \delta$ are any real numbers. Then there exist integers p, q, not both zero, for which simultaneously

$$|\xi| = |\alpha p + \beta q| \leq \sqrt{|\Delta|}, \quad |\eta| = |\gamma p + \delta q| \leq \sqrt{|\Delta|}.$$

Proof. Take for our M-set the parallelogram in Figure 6.2 bounded by the four lines

$$\alpha x + \beta y = \pm k, \quad \gamma x + \delta y = \pm l, \quad \text{where } k > 0,$$

and with determinant $\Delta = \alpha\delta - \beta\gamma \neq 0$. The area of this parallelogram is twice the area of the triangle $A : (x_1, y_1), B : (x_2, y_2), C : (x_3, y_3)$. Since the parallelogram is symmetric, $(x_3, y_3) = (-x_1, -y_1)$.

From analytic geometry, we know how to state the area of triangle ABC in determinant form. It is

$$\text{Area } ABC = \frac{1}{2} \begin{vmatrix} x_1 & y_1 & 1 \\ x_2 & y_2 & 1 \\ x_3 & y_3 & 1 \end{vmatrix} = \frac{1}{2} \begin{vmatrix} x_1 & y_1 & 1 \\ x_2 & y_2 & 1 \\ -x_1 & -y_1 & 1 \end{vmatrix}.$$

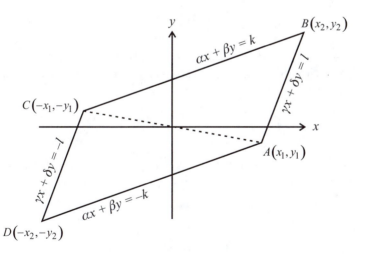

Figure 6.2

Parallelogram for proof of Theorem 6.2

By expanding the last determinant with respect to the third column, we obtain

$$\text{Area } ABC = |x_1 y_2 - x_2 y_1|. \tag{6.1}$$

What are the coordinates of A and B? We find them by solving the systems

$$\begin{aligned}
\alpha x + \beta y &= -k \\
\gamma x + \delta y &= l
\end{aligned} \quad \text{and} \quad \begin{aligned}
\alpha x + \beta y &= k \\
\gamma x + \delta y &= l,
\end{aligned}$$

respectively. The respective solutions of these two linear systems are

$$\begin{aligned}
x_1 &= \frac{-1}{\Delta}(\delta k + \beta l) \\
y_1 &= \frac{1}{\Delta}(\alpha l + \gamma k)
\end{aligned} \quad \text{and} \quad \begin{aligned}
x_2 &= \frac{1}{\Delta}(\delta k - \beta l) \\
y_2 &= \frac{1}{\Delta}(\alpha l - \gamma k).
\end{aligned}$$

By substituting these values into (6.1), we obtain the area of the triangle,

$$\text{Area } ABC = \frac{2k|l|\,|\alpha\delta - \beta\gamma|}{\Delta^2} = \frac{2k|l|}{|\Delta|} = 2k\left|\frac{l}{\Delta}\right|,$$

which when doubled becomes the area of the parallelogram in Figure 6.2,

$$A = 4k\left|\frac{l}{\Delta}\right|.$$

Now choose $l = \Delta/k$, where k is any assigned positive number, so that $A = 4$. Then, by Minkowski's Theorem, there exist integers p', q', not both zero, such that

$$|\alpha p' + \beta q'| \le k, \qquad |\gamma p' + \delta q'| \le \frac{|\Delta|}{k}.$$

In particular, if $k = \sqrt{|\Delta|}$, then integers p, q, not both zero, exist, such that simultaneously

$$|\alpha p + \beta q| \le \sqrt{|\Delta|}, \qquad |\gamma p + \delta q| \le \sqrt{|\Delta|}.$$

This is precisely the statement of Theorem 6.2.

Take note of the condition on the determinant. For if $\Delta = 0$, then $\alpha/\beta = \gamma/\delta$ and $\xi = (\beta/\delta)\eta = \eta$, a situation of no interest.

Problem Set for Section 6.2

1. Apply Minkowski's First Theorem to the linear forms $\xi = 230x + 201y$ and $\eta = 459x - 400y$. Calculate actual values of $x = p$ and $y = q$ that satisfy the theorem.

2. Repeat Problem 1 for $\xi = 230x + 201y$ and $\eta = 459x + 400y$.

6.3   Minkowski's Second Theorem

Here is an application of Minkowski's First Theorem. Again, we require the determinants to be nonzero.

Theorem 6.3 (Minkowski's Second Theorem).   *Given two linear forms*

$$\xi = \alpha x + \beta y, \qquad \text{determinant } \Delta = \alpha\delta - \beta\gamma \neq 0,$$
$$\eta = \gamma x + \delta y,$$

where $\alpha, \beta, \gamma, \delta$ are any real numbers. Then there exist integers p, q, not both zero, for which

$$|\xi\eta| = |\alpha p + \beta q| \cdot |\gamma p + \delta q| \leq \frac{1}{2}|\Delta|.$$

Proof.   We add and subtract the two linear forms given in the statement of the theorem. This produces

$$\xi + \eta = (\alpha + \gamma)x + (\beta + \delta)y,$$
$$\xi - \eta = (\alpha - \gamma)x + (\beta - \delta)y.$$

Next, we find the determinant D of this new linear system in the usual way:

$$D = (\alpha + \gamma)(\beta - \delta) - (\alpha - \gamma)(\beta + \delta)$$
$$= -2(\alpha\delta - \beta\gamma) = -2\Delta \neq 0.$$

Thus we have $|D| = 2|\Delta|$.

By Theorem 6.2, it is possible to find two integers p, q, not both zero, such that simultaneously

$$|\xi + \eta| \leq \sqrt{|D|} = \sqrt{2|\Delta|},$$
$$|\xi - \eta| \leq \sqrt{|D|} = \sqrt{2|\Delta|}.$$

However, $|\xi| + |\eta|$ is the larger of the numbers $|\xi + \eta|$ and $|\xi - \eta|$; that is,

$$|\xi| + |\eta| = \max\left\{|\xi + \eta|, |\xi - \eta|\right\}.$$

With both $|\xi + \eta|$ and $|\xi - \eta|$ less than or equal to $\sqrt{2|\Delta|}$, it follows that

$$|\xi| + |\eta| \leq \sqrt{2|\Delta|}. \tag{6.2}$$

For our last step, we need the *arithmetic–geometric mean inequality*. This states that, if a and b are real positive numbers, then

$$\sqrt{ab} \leq \frac{a+b}{2}, \quad \text{or} \quad ab \leq \left(\frac{a+b}{2}\right)^2.$$

This is true because $0 \leq \left(\sqrt{a} - \sqrt{b}\right)^2 = a + b - 2\sqrt{ab}$, so $2\sqrt{ab} \leq a + b$; thus $\sqrt{ab} \leq (a + b)/2$. By this argument,

$$|\xi\eta| = |\xi| \cdot |\eta| \leq \left(\frac{|\xi| + |\eta|}{2}\right)^2 \leq \left(\frac{\sqrt{2|\Delta|}}{2}\right)^2 = \frac{1}{2}|\Delta|,$$

which proves Theorem 6.3.

Remarks. Actually, the quantity $\frac{1}{2}$ in Theorem 6.3 is not the "best possible" constant. It can be proved that there exist integers p, q, not both zero, for which

$$|\xi\eta| \leq \frac{|\Delta|}{\sqrt{5}}.$$

We consider $\sqrt{5}$ the "best possible" constant, in the sense that if any larger number replaces it, the theorem will not hold. For the (rather difficult) proof of this refinement, see Hardy and Wright [**1,** Ch. 24, Th. 454].

Problem for Section 6.3

1. Apply Minkowski's Second Theorem to the linear systems $\xi = x - \pi y$ and $\eta = x - ey$, where $\pi = 3.14159\ldots$ and $e = 2.71828\ldots$.

6.4   Approximating Irrational Numbers

As an application of Minkowski's Second Theorem, we will state and prove yet another theorem. This is the one referred to in Chapter 5 pertaining to the approximation of irrational numbers by rationals.

Theorem 6.4. *Given any real irrational number α, there exist rational fractions p/q, with arbitrarily large denominators q, such that $|(p/q) - \alpha| \leq 1/2q^2$.*

Proof. Consider the two linear forms

$$\xi = t(x - \alpha y), \quad \eta = \frac{y}{t}, \quad \text{for } t \neq 0,$$

where α is given and t can be as large as we please. The determinant of these forms is

$$\Delta = \begin{vmatrix} -t\alpha & t \\ 1/t & 0 \end{vmatrix} = -1 \neq 0, \quad \text{where } |\Delta| = 1.$$

By Theorem 6.3, integers p, q exist, not both zero, such that

$$|\xi\eta| = |t(p - \alpha q)| \cdot \left|\frac{q}{t}\right| \leq \frac{|-1|}{2} = \frac{1}{2} \tag{6.3},$$

which is the same as the inequality

$$|p - \alpha q| \cdot |q| \leq \frac{1}{2}. \tag{6.4}$$

By invoking inequality (6.2) from our proof of Theorem 6.3, we also have

$$|\xi| + |\eta| = |t(p - \alpha q)| + \left|\frac{q}{t}\right| \leq \sqrt{2|\Delta|} = \sqrt{2}. \tag{6.5}$$

It follows from (6.5) that

$$|p - \alpha q| \leq \frac{\sqrt{2}}{t}. \tag{6.6}$$

We claim that, by taking t sufficiently large, we can make q larger than any integer N. Consider for each integer $q \neq 0$,

$$\min_{\text{integer } p} |p - \alpha q| = m(q).$$

Since α is irrational, $m(q) > 0$ for every integer q. And for $q = 0$,

$$m(0) = \min_{\text{integer } p \neq 0} |p| = 1.$$

Now, define

$$m = \min \big(m(0), m(1), \ldots, m(N)\big),$$

and choose t so large that $\sqrt{2}/t < m$. Then (6.6) states that $|p - \alpha q| < m$, which—since $m \leq$ any $m(q)$, where $q \leq N$—is not satisfied for integers

$q \le N$; so q is greater than N, chosen arbitrarily. Dividing (6.4) by q^2, we get the desired inequality

$$\left| \frac{p}{q} - \alpha \right| \le \frac{1}{2q^2}.$$

Remarks. Theorem 6.4 tells us that we can approximate any irrational number α by a rational fraction p/q, where $q \ne 0$, to the degree of accuracy indicated. Again, this is not the "best possible" result. A famous theorem due to Hurwitz [**3**] states that any irrational number α has an infinity of rational approximations p/q such that

$$\left| \frac{p}{q} - \alpha \right| \le \frac{1}{\sqrt{5}q^2}.$$

Here, the number $\sqrt{5}$ is the "best possible" because if any larger number replaces it the theorem will not hold [**1**, Ch. 11, Th. 194].

6.5 Minkowski's Third Theorem

Finally, for completeness, we state Minkowski's Third Theorem. It is much more difficult to prove than the others, hence we will not attempt its proof here.

Theorem 6.5 (Minkowski's Third Theorem). *If ξ, η, and Δ are defined as in Theorem 6.3, then to every real ζ and σ there corresponds a pair of integers p, q for which $\left| (\xi - \zeta)(\eta - \sigma) \right| \le \frac{1}{4} |\Delta|$.*

Remarks. Suppose that in Theorem 6.5 we let $\xi = p - \alpha q$, $\eta = q$, $\zeta = c$, and $\sigma = 0$, where α is irrational. Substituting, we obtain the inequalities

$$\left| (p - \alpha q - c)q \right| \le \frac{1}{4},$$

or

$$|p - \alpha q - c| \le \frac{1}{4|q|},$$

provided that $q \ne 0$. For a proof of Theorem 6.5 and a discussion of some interesting questions that arise from the inequalities, see Hardy and Wright [**1**, Ch. 24, Sect. 6–8].

6.6 Simultaneous Diophantine Approximations

We conclude this chapter with an example of how Minkowski's Theorem can be applied to *simultaneous Diophantine approximation*, which is the simultaneous approximation of n irrational numbers $\alpha_1, \alpha_2, \ldots, \alpha_n$ by n rational fractions, each having the same denominator.

Theorem 6.6. *Let $\alpha_1, \alpha_2, \ldots, \alpha_n$ be any n irrational numbers. Then there exist infinitely many sets of integers $p_1, p_2, \ldots, p_n, p$, with $p \geq 1$, such that, simultaneously,*

$$\left| \alpha_1 - \frac{p_1}{p} \right| < \frac{1}{p^{(n+1)/n}},$$

$$\left| \alpha_2 - \frac{p_2}{p} \right| < \frac{1}{p^{(n+1)/n}},$$

$$\vdots$$

$$\left| \alpha_n - \frac{p_n}{p} \right| < \frac{1}{p^{(n+1)/n}}.$$

Proof. Let s be any number less than 1. Consider the region K in $\mathbb{R}^{n+1}$ consisting of all points $(x_1, x_2, \ldots, x_n, y)$ satisfying the inequalities

$$|x_i - \alpha_i y| \leq s, \quad i = 1, 2, \ldots, n, \quad |y| \leq s^{-n}.$$

This region K is a parallelopiped centered at the origin; it is closed, symmetric with respect to the origin, and convex.

We claim its volume is 2^{n+1}. To see this, make a linear mapping from $x_1, x_2, \ldots, x_n, y$-space to $u_1, u_2, \ldots, u_n, v$-space as follows:

$$u_i = \frac{1}{s}(x_i - \alpha_i y), \quad v = s^n y.$$

This mapping carries the region K in $x_1, x_2, \ldots, x_n, y$-space into H in $u_1, u_2, \ldots, u_n, v$-space defined by the inequalities

$$|u_i| \leq 1, \quad |v| \leq 1.$$

Clearly, H is a cube in $\mathbb{R}^{n+1}$ with edge length 2. So vol$(H) = 2^{n+1}$. This map is volume preserving, as can be seen by verifying that its Jacobian determinant is 1. It follows that also vol$(K) = 2^{n+1}$.

Thus, by Minkowski's Fundamental Theorem, the region K contains a lattice point $(p_1, p_2, \ldots, p_n, p)$ other than the origin $O : (0, 0, \ldots, 0)$. So

$$|p_i - \alpha_i p| \leq s, \quad i = 1, 2, \ldots, n, \quad |p| \leq s^{-n}. \tag{6.7}$$

We may assume that $p \geq 0$; if not, just change the sign of all p_i and of p. We claim that p is positive. For, if $p = 0$, inequalities (6.7) would assert that $|p_i| \leq s$; since s was chosen < 1 and p_i are integers, they would all be zero. Then $(p_1, p_2, \ldots, p_n, p) = (0, 0, \ldots, 0)$, contrary to our choice.

Dividing the first n inequalities in (6.7) by p, we get

$$\left| \frac{p_i}{p} - \alpha_i \right| \leq \frac{s}{p}.$$

From the last inequality in (6.7), we deduce that $s \leq p^{-1/n}$. Setting this into the inequality above, we obtain

$$\left| \frac{p_i}{p} - \alpha_i \right| \leq \frac{1}{p^{1+1/n}}, \quad \text{for } i = 1, 2, \ldots, n, \tag{6.8}$$

as desired.

To show that these inequalities have infinitely many solutions, take any finite set of solutions of (6.8) and choose s so small that $|p_i - \alpha_i p| > s$ for all solutions in this finite set. Then any solution of (6.7) is clearly different from all these solutions of (6.8).

Reading Assignment for Chapter 6

Minkowski's Fundamental Theorem implies that if a square of side 2 is superimposed on the lattice Λ in such a way that its center coincides with a lattice point, then there is bound to be another lattice point inside the square or on its boundary. Study the proof given in Hilbert and Cohn-Vossen [2] and notice the influence of Minkowski's intuitive ideas.

References

1. G. H. Hardy and E. M. Wright, *An Introduction to the Theory of Numbers*, 5th ed. (Oxford: Oxford University Press, 1983).
2. David Hilbert and S. Cohn-Vossen, *Geometry and the Imagination*, translated by P. Nemenyi (New York: Chelsea, 1952), 41.

3. A. Hurwitz, "Über die angenäherte Darstellung der Irrationalzahlen durch rationale Brüche," *Mathematische Annalen* 39 (1891):279–84.

4. Hermann Minkowski, *Ausgewahlte Arbeiten zur Zahlentheorie und zur Geometrie. Mit D. Hilbert's Gedachtnisre auf H. Minkowski (Göttingen, 1909)* [*Selected Papers on Number Theory and Geometry. With D. Hilbert's Commemorative Address in Honor of H. Minkowski*], Teubner-Archiv zur Mathematik, Vol. 12, E. Kratzel and B. Weissbach, eds. (Leipzig: Teubner, 1989).

7

Linear Transformations
and Integral Lattices

7.1 Linear Transformations

Those of you who read the optional Section 5.4 are probably already familiar with much of the material we are about to discuss. For those readers who are encountering the geometry of numbers for the first time, we hope that you will be stimulated to understand the proofs of more difficult theorems or even go on to do independent study in the literature. A prerequisite for more advanced work of this sort is some knowledge of *linear transformations*. In this chapter, we will get a sense of how linear transformations apply to the study of integral lattices. We begin by defining the procedure of linear transformation.

Any point (x, y) in the xy-plane can be *transformed into* the point (x', y') by a *linear transformation*, denoted by T, if (x', y') can be expressed in terms of (x, y) by a pair of linear equations. Symbolically, we write this as

$$T : \begin{array}{l} x' = ax + by, \\ y' = cx + dy, \end{array} \qquad \text{with } \Delta = ad - bc \neq 0,$$

where the coefficients a, b, c, d are given real constants. We assume that the *determinant* $\Delta = ad - bc$ of the transformation is not zero. When (x, y) has been transformed into (x', y') under T, we write this replacement symbolically as

$$T : (x, y) \to (x', y').$$

Besides assuming a nonzero determinant, a second, crucial, assumption we make is that both (x, y) and (x', y') are plotted with respect to the original xy-axes. Why is this important? Because otherwise we can quickly become confused. For instance, consider the transformation

$$T_1 : \quad \begin{aligned} x' &= x + y, \\ y' &= x - y, \end{aligned} \qquad \text{with } \Delta = -1 - 1 = -2.$$

First adding, $x' + y'$, and then subtracting, $x' - y'$, we see that we can express x and y as

$$x = \frac{1}{2}(x' + y'), \qquad y = \frac{1}{2}(x' - y').$$

Under T_1 the x-axis, $y = 0$, has been transformed into the line $x' - y' = 0$, which is no longer the x-axis. Do not think of T_1 as changing the x-axis into the line $y' = x'$! Rather, say that the line $y = 0$ originally coincided with the x-axis but now coincides with the line $y' = x'$. Compare both lines in Figure 7.1, where we also see how under T_1 the point $(x, y) = (2, 1)$ is transformed into the point $(3, 1)$. Again, both points are plotted with respect to the original axes.

The *inverse transformation* of T, denoted T^{-1}, is a function that undoes the transformation T. For example, consider our original transformation equations,

$$T : \quad \begin{aligned} x' &= ax + by, \\ y' &= cx + dy, \end{aligned} \qquad \text{with } \Delta = ad - bc \neq 0.$$

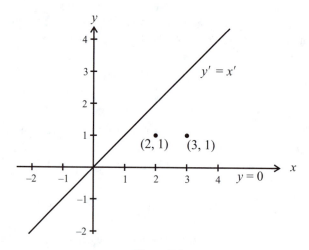

Figure 7.1

The linear transformation $T_1 : x' = x + y, \ y' = x - y$.

If we solve for x and y in terms of x' and y', we get the inverse transformation

$$T^{-1}: \begin{array}{l} x = a_1 x' + b_1 y', \\ y = c_1 x' + d_1 y', \end{array} \qquad \text{with } \Delta_1 = a_1 d_1 - b_1 c_1 \neq 0,$$

where $a_1 = d/\Delta$, $b_1 = -b/\Delta$, $c_1 = -c/\Delta$, and $d_1 = a/\Delta$. The determinant of the inverse transformation T^{-1} is

$$\Delta_1 = a_1 d_1 - b_1 c_1 = \frac{ad - bc}{\Delta^2} = \frac{1}{\Delta} \neq 0,$$

the reciprocal of the determinant, Δ, of T.

This shows two things: First of all, transformations with zero determinant do not have inverses, so we are restricting our attention to invertible ones. Secondly, we will meet special transformations with determinant $\Delta = \pm 1$, and, for them, $\Delta_1 = \pm 1$ as well.

Linear transformations have many useful properties. Here are five properties that are particularly powerful. Under a linear transformation T:

1. Points and lines are transformed into, respectively, points and lines.
2. Conic sections (circles, ellipses, etc.) are transformed into conic sections.
3. A point dividing a line segment in a given ratio is transformed into a point dividing the transformed line segment in the same ratio. (One consequence for us is that M-sets are thus transformed into M-sets.)
4. When a, b, c, d are given integers in T and its determinant $\Delta = ad - bc = \pm 1$, then if x and y are integers, so are x' and y', and vice versa. In other words, a linear transformation T with integer coefficients and determinant ± 1 maps lattice points into lattice points; the same holds for the inverse T^{-1} of T.
5. Areas are invariant under linear transformations of determinant $\Delta = \pm 1$.

You will be exploring properties (1) through (3) in Problem 1, while property (4) is applied in Problem 4. Property (5), however, is harder to prove and involves calculus; so we shall simply accept the statement as true. After all, it is easy to see that under a linear transformation of determinant $\Delta = \pm 1$ the areas of triangles and rectangles are invariant, and the concept of area in elementary calculus is based on the limit of sums of approximating rectangles.

Problem Set for Section 7.1

1. For a linear transformation T:

   a. Prove property (1).

   b. Prove property (2).

   c. Prove property (3).

2. Consider the rectangle $(0,0)$, $(1,0)$, $(1,2)$, $(0,2)$.

   a. Transform this rectangle, using the transformation $T : x' = x$, $y' = x + y$, and graph its image.

   b. Calculate the areas of the original rectangle and of its image under T.

3. Consider the triangle $(0,0)$, $(10,0)$, $(10,10)$. Under the transformation $T : x = x' + y'$, $y = x' + 2y'$, it is transformed into a new triangle.

   a. Find the determinant of T.

   b. Compare the areas of the original and the transformed triangles.

4. Under the transformation $T : x' = 2x + 3y, y' = 4x + 6y$, the vertices $(0,0)$, $(1,0)$, $(1,1)$, $(0,1)$ of a square are transformed into the points $(0,0)$, $(2,4)$, $(5,10)$, $(3,6)$. Thus, under T,

$$(x, y) = (0, 1) \rightarrow (x', y') = (3, 6).$$

   a. Graph the image of the given square under T.

   b. Is there an inverse transformation T^{-1} that maps $(x', y') = (3,6) \rightarrow (x, y) = (0, 1)$? Explain.

7.2 The General Lattice

In Chapter 1, we introduced the fundamental lattice of lines L and the fundamental point-lattice Λ, which is determined by the intersection of the lines of L. Because lattice points are our main concern, we will adopt the shorthand "lattice" to mean "point-lattices" in later discussions. Before getting that familiar, however, let's stop and examine the general concept of a lattice.

In the xy-plane, select a point O as the origin and choose two points P and Q, such that O, P, Q are not collinear, as depicted in Figure 7.2. Draw a line ℓ through O and P, and draw another line ℓ' through O and Q. Along ℓ, measure off equal intervals of length $a = |\overline{OP}|$, and do the same for length $b = |\overline{OQ}|$ along ℓ'. Now draw two systems of lines: first, lines parallel to ℓ' through the equally spaced points on ℓ; second, lines

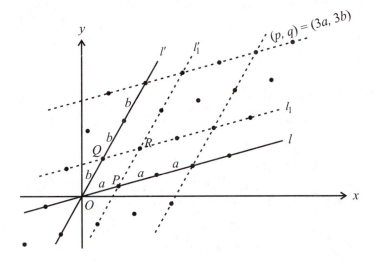

Figure 7.2
Constructing a point-lattice.

parallel to ℓ through the equally spaced points on ℓ'. These two systems of lines form a *lattice*. The points of intersection of the lines are called *lattice points*, and the totality of such points is a *point-lattice*.

What have we constructed? As Figure 7.2 shows, we have divided the plane into an infinite number of parallelograms, all congruent and similarly placed to the parallelogram $OPRQ$. We say that such a lattice is *based* on the three points O, P, Q. Actually, it is more precise to say that the lattice is based on the two directed line segments (or *vectors*) $\overrightarrow{OP}$ and $\overrightarrow{OQ}$, for given only these two directed line segments we could still reconstruct the lattice.

When a lattice is based on the three points O, P, Q as described, then $OPRQ$ is called the *fundamental parallelogram* of the lattice. A parallelogram whose vertices are lattice points, but which has no other lattice points on its boundary or in its interior, is called a *primitive parallelogram*. For instance, in Figure 7.2, $OPRQ$ is primitive.

It is important to notice that two distinctly different lattices can produce the same point-lattice. In Figure 7.3, for example, the lattice based on $\overrightarrow{OP}$ and $\overrightarrow{OS}$ produces the same system of lattice points as the lattice based on $\overrightarrow{OP}$ and $\overrightarrow{OQ}$. Two lattices that produce the same point-lattice are said to be *equivalent*.

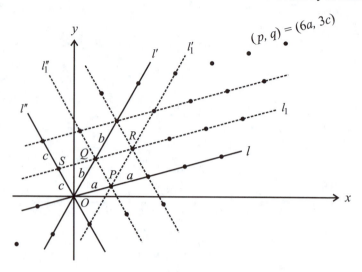

Figure 7.3
Equivalent point-lattices.

7.3 Properties of the Fundamental Lattice Λ

Now we can put some key terms into context. Our *fundamental point-lattice* Λ is *based* on the three points $O : (0,0)$, $P : (1,0)$, and $Q : (0,1)$. Its *fundamental parallelogram* is the *unit square OPRQ*, which is *primitive*. Each lattice point of Λ has *integer coordinates* (p,q).

Now, let (p,q) represent any *lattice point* of Λ. We can apply to (p,q) the *transformation*

$$T: \quad \begin{aligned} p' &= ap + bq, \\ q' &= cp + dq, \end{aligned} \quad \text{with } \Delta = ad - bc \neq 0, \quad (7.1)$$

where a, b, c, d are given integers. Any lattice point of Λ is transformed into a lattice point (p', q') of Λ', since both p' and q' clearly are integers. In other words, all the lattice points (p', q') of Λ' are points of Λ.

This lattice transformation raises an interesting question:

Will the lattice points of Λ' merely reproduce, in some order, the lattice points of Λ, or will Λ' be a subset of the points of Λ?

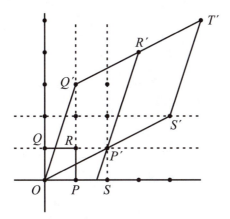

Figure 7.4

Transformation of Λ into a subset Λ'.

For example, suppose the lattice points of Λ are subjected to the transformation

$$T_2 : \begin{array}{l} p' = 2p + q, \\ q' = p + 3q, \end{array} \quad \text{with } \Delta = (2 \cdot 3) - (1 \cdot 1) = 6 - 1 = 5.$$

Consulting Figure 7.4, we see the emergence of a new lattice as

$$O : (0,0) \rightarrow O' : (0,0) = 0 : (0,0)$$
$$P : (1,0) \rightarrow P' : (2,1)$$
$$R : (1,1) \rightarrow R' : (3,4)$$
$$Q : (0,1) \rightarrow Q' : (1,3).$$

The lattice Λ, based on $\overrightarrow{OP}$ and $\overrightarrow{OQ}$, is thus transformed into a new lattice Λ', based on $\overrightarrow{OP'}$ and $\overrightarrow{OQ'}$, all of whose points are still points of Λ. For instance, the primitive parallelogram $OPRQ$ is transformed into a nonprimitive parallelogram $OP'R'Q'$ of Λ'. The adjacent square $PSP'R$ of Λ is transformed into the adjacent parallelogram $P'S'T'R'$ of Λ', and so on. Although all lattice points of Λ' coincide with points of Λ, some points of Λ are not points of Λ'. So Λ' is a subset of Λ.

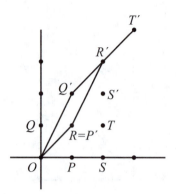

Figure 7.5

Λ' and Λ are equivalent.

Here is another example of a transformation—but with a different outcome. Referring to Figure 7.5, consider the transformation

$$T_3: \quad \begin{aligned} p' &= p + q, \\ q' &= p + 2q, \end{aligned} \qquad \text{with } \Delta = (1 \cdot 2) - (1 \cdot 1) = 2 - 1 = 1.$$

Observe that

$$O : (0,0) \to O' : (0,0) = O : (0,0)$$
$$P : (1,0) \to P' : (1,1) = R : (1,1)$$
$$R : (1,1) \to R' : (2,3)$$
$$Q : (0,1) \to Q' : (1,2).$$

What is happening here? This time the primitive parallelogram $OPRQ$ is transformed into the primitive parallelogram $ORR'Q'$, while adjacent squares are transformed into adjacent parallelograms. The lattice points of Λ' simply reproduce the lattice points of Λ, so all points of Λ are at the same time points of Λ'. We say that, under T_3, the lattice Λ has been transformed "into itself." In other words, Λ' is *equivalent* to Λ.

These two different outcomes raise a new question:

Why does one linear transformation transform Λ into itself, while another does not?

You have probably already guessed that the answer has to do with the determinant, specifically, whether $|\Delta|$ equals 1 or not.

To explore the role of the determinant, let's return to the transformation (7.1):

$$T: \quad \begin{aligned} p' &= ap + bq, \\ q' &= cp + dq, \end{aligned} \quad \text{with } a, b, c, d \text{ integers, } \Delta \neq 0.$$

We know that every point (p, q) of Λ is transformed under T into another lattice point (p', q') of Λ. Solving (7.1) for p and q in terms of p' and q', we obtain the inverse transformation

$$T^{-1}: \quad \begin{aligned} p &= \frac{dp' - bq'}{\Delta}, \\ q &= \frac{aq' - cp'}{\Delta}. \end{aligned} \tag{7.2}$$

Substituting $\Delta = ad - bc = \pm 1$ into (7.2) then gives us

$$T^{-1}: \quad \begin{aligned} p &= \pm(dp' - bq'), \\ q &= \pm(aq' - cp'). \end{aligned}$$

This ensures that any integral values of p' and q' will give integral values of p and q. Thus every lattice point (p', q') will correspond to a lattice point (p, q) of Λ, so Λ is transformed into itself.

Let's take a closer look at the conditions that must be met for Λ to be transformed into itself. Specifically, regarding the determinant Δ:

In order for the transformation T to carry Λ into an equivalent lattice Λ', is it absolutely necessary that $\Delta = ab - bc = \pm 1$?

Suppose we know only that $\Delta \neq 0$ and that T in (7.1) transforms Λ into itself. This means that not only does T map lattice points into lattice points, but every lattice point of Λ' is the image under T of some lattice point of Λ. In other words, the inverse transformation T^{-1} given by (7.2) maps any lattice point (p', q') of Λ' to a lattice point of Λ. For instance, T^{-1} must map $(p', q') = (1, 0)$ of Λ' into a lattice point (p, q) of Λ.

This equivalence implies that equations (7.2) with $p' = 1$, $q' = 0$ must yield integers. That is, both

$$p = \frac{dp' - bq'}{\Delta} = \frac{(d \cdot 1) - (b \cdot 0)}{\Delta} = \frac{d}{\Delta}$$

and

$$q = \frac{aq' - cp'}{\Delta} = \frac{(a \cdot 0) - (c \cdot 1)}{\Delta} = -\frac{c}{\Delta}$$

must be integers. We express this symbolically by the exact divisor nota-tion: $\Delta | d$ and $\Delta | c$. Moreover, T^{-1} must also transform $(p', q') = (0, 1)$ of Λ' into a lattice point (p, q) of Λ, similarly implying that $\Delta | b$ and $\Delta | a$.

Let's suppose that, indeed, Δ divides each of the integers a, b, c, d. Then $\Delta^2 | ad$ and $\Delta^2 | bc$, hence

$$\Delta^2 | ad - bc = \Delta.$$

In short, $\Delta^2 | \Delta$. If this is so, then

$$\frac{\Delta}{\Delta^2} = \frac{1}{\Delta},$$

which must be an integer. Thus $\Delta = \pm 1$ necessarily.

We have proved the following theorem.

Theorem 7.1. *If a, b, c, d are integers, a necessary and sufficient condi-tion that the transformation $T : p' = ap + bq$, $q' = cp + dq$, transform Λ into itself is that $\Delta = ad - bc = \pm 1$.*

For another view of Theorem 7.1, let's consider the transformation in terms of area. Consider a point-lattice Λ' based on the three non-collinear points $O : (0, 0)$, $P' : (a, c)$, and $Q' : (b, d)$ of Λ, as in Figures 7.4 and 7.5. All the lattice points (p', q') of Λ' are given by the equations

$$\begin{aligned} p' &= pa + qb, \\ q' &= pc + qd, \end{aligned} \tag{7.3}$$

where p and q can take arbitrary integer values. By the way, readers familiar with vectors will observe that equations (7.3) correspond to the single vector equation

$$(p', q') = p(a, c) + q(b, d),$$

which gives all linear combinations with integer coefficients of the two linearly independent vectors (a, c) and (b, d).

Let's take Figure 7.4 as an illustration of how area is changed under a linear transformation. To reproduce all the lattice points of Λ' we simply use $O : (0, 0)$, $P' : (2, 1)$, $Q' : (1, 3)$ as the base and apply the equations

$$\begin{aligned} p' &= 2p + q, \\ q' &= p + 3q, \end{aligned} \qquad \text{with } \Delta = (2 \cdot 3) - (1 \cdot 1) = 5.$$

In turn, letting

$$(p, q) = (1, 1), (2, 0), (2, 1),$$

we immediately get

$$(p', q') = (3, 4), (4, 2), (5, 5)$$

—namely, the points R', S', and T'.

Now the area of the parallelogram $OP'R'Q'$ is twice the area of the triangle $OP'Q'$. Symbolically, we write this as

$$\text{Area } OP'R'Q' = 2 \cdot \frac{1}{2} \begin{vmatrix} 0 & 0 & 1 \\ a & b & 1 \\ c & d & 1 \end{vmatrix} = \begin{vmatrix} a & b \\ c & d \end{vmatrix} = ad - bc = \Delta.$$

Notice the orientation of the points O, P', Q' in Figure 7.4. Had we applied this formula to O, Q', P', oriented in the opposite direction, we would have gotten $-(ad - bc)$. In either case, the area is $|ad - bc|$.

This interpretation gives us another way to describe the equivalent transformation of Theorem 7.1. We state it in the following theorem.

Theorem 7.2. *For the lattice Λ' based on $\overrightarrow{OP'}$ and $\overrightarrow{OQ'}$ to be equivalent to Λ, a necessary and sufficient condition is that the parallelogram defined by $\overrightarrow{OP'}$ and $\overrightarrow{OQ'}$ have unit area.*

Problem Set for Section 7.3

1. Prove that the transformation T given in (7.1) cannot transform two different points (p_1, q_1) and (p_2, q_2) of Λ into the same point (p', q') of Λ'.

2. Give two examples of a transformation T that will transform Λ into itself. For each, make sketches similar to Figure 7.4.

3. Is it possible to construct a parallelogram $OPRQ$, where O, P, R, Q are lattice points of Λ, that has an area $A < 1$? Explain.

4. *Reading assignment.* Study "The Farey Series," Essay 5 in Ross Honsberger's *Ingenuity in Mathematics*, New Mathematical Library Series, Vol. 23 (New York: Random House, 1970), 24–37.

7.4   Visible Points

Viewed from the origin $O : (0, 0)$, a point P of Λ is said to be *visible* if no other lattice point of Λ is blocking our view of P—that is, when no lattice point of Λ lies between O and P on the line segment $\overline{OP}$. In Chapter 1, we proved the necessary and sufficient condition for such visibility: the point $P : (p, q)$ will be visible from $O : (0, 0)$ if p and q are relatively prime. This concept of the visible point gives us a new way to locate lattice points, as stated in the following theorem.

Theorem 7.3. *If P and Q are two visible points of Λ, and if the area of the parallelogram K based on $\overrightarrow{OP}$ and $\overrightarrow{OQ}$ is equal to δ, then:*

1. *if $\delta = 1$, no point of Λ exists inside K;*
2. *if $\delta > 1$, at least one point of Λ exists inside K.*

Proof. The proof of part (1) follows immediately from Theorem 7.2. Suppose, on the contrary, $\delta = 1$ and a lattice point of Λ does exist inside K. Then the lattice Λ' based on $\overrightarrow{OP}$ and $\overrightarrow{OQ}$ would not be equivalent to Λ, contrary to the statement of Theorem 7.2. The situation for $\delta = 1$ is illustrated in Figure 7.6a.

On the other hand, to prove part (2), suppose $\delta > 1$, so Λ and Λ' are not equivalent. Then at least one lattice point S' must exist inside or on the boundary (excluding the vertices) of the fundamental parallelogram

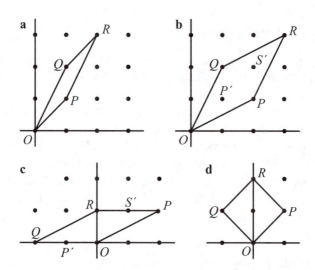

Figure 7.6
Parallelograms based on visible points P and Q. In (a), $\delta = 1$; in (b), (c), (d), $\delta > 1$.

K. As Figures 7.6b and 7.6c show, the symmetry of the lattice Λ means that in general a second lattice point P' must exist in K, since every point S' in triangle PQR has a corresponding point P' in the congruent triangle OPQ. The two points S' and P' might even coincide at the point of intersection of the diagonals OR and PQ of K, as in Figure 7.6d.

8
Geometric Interpretations of Quadratic Forms

8.1 Quadratic Representation

The study of *quadratic forms in two or more variables* takes us into some of the most advanced parts of the theory of numbers. The whole third volume of Dickson's *History of the Theory of Numbers* [**6**], for example, is devoted to the subject—and Dickson stopped with the 1920s! It was Joseph Louis Lagrange (1736–1813), the foremost mathematician of the eighteenth century (rivalled only by Euler), who published the first proof that every positive integer can be expressed as the sum of at most four squares (see Section 8.6). Lagrange's theory of quadratic forms, first developed in 1773, was later simplified and extended by his younger contemporaries Legendre and Gauss.

What is a *binary quadratic form*? It is an expression

$$f(x,y) = ax^2 + 2bxy + cy^2, \qquad (8.1)$$

where $a, 2b, c$ are integers and x and y are two real variables (hence the adjective "binary"). We are interested in the behavior of such a quadratic form as x and y are replaced by pairs of integers. The classic problem related to quadratic forms is that of representation:

> *Given a particular quadratic form, what integers n can be represented by $f(x,y)$, as x and y range over all possible integer values?*

In other words:

> *What integers n can be expressed in the form $n = ax^2 + 2bxy + cy^2$?*

It is known what integers n can be represented by special quadratic forms, such as $n = x^2 + y^2$, $n = x^2 + 2y^2$, $n = x^2 + 3y^2, \ldots$. However, no theory tells us what integers n can be represented by an arbitrary quadratic form.

Nonetheless, many questions concerning quadratic forms are approachable. One such question can be related to the geometry of numbers and brings Minkowski's Fundamental Theorem into play. It is this:

What is the smallest number M such that $|f(x, y)| \geq M$ for integral values of x and y?

If we actually identify this smallest number M, then we can say we have found the *lower bound* or *minimum* of $|f(x, y)|$. Successfully pinpointing a minimum is sometimes too ambitious a project, however, in which case we will settle for finding some nontrivial upper estimate of M.

In this chapter we will look at only one type of quadratic form, namely, those with values $f(x, y) \geq 0$. (Note that $f(0, 0) = 0$.) Such a form is called a *positive definite quadratic form*. To ensure that all nonzero values are positive, we must add some conditions to (8.1). In particular, we will need to specify that

$$f(x, y) = ax^2 + 2bxy + cy^2, \tag{8.2}$$

where a, b, c are integers, have *discriminant* $d = ac - b^2 > 0$. The positivity of the discriminant limits the set of values that the form can take. To see this, note that we assume at the outset that $a > 0$, so we must have $c > 0$ as well. After multiplying $f(x, y)$ by a, we complete the square to obtain

$$af(x, y) = (ax + by)^2 + (ac - b^2)y^2.$$

Since $d = ac - b^2 > 0$, the terms on the right will always be positive for any integer pairs (x, y) other than $(0, 0)$. Hence, we see that such a form can represent only zero or positive numbers.

8.2 An Upper Bound for the Minimum Positive Value

How do we look for the minimum or lower bound $|f(x, y)| \geq M$ of a positive definite quadratic form? Let's quickly review some techniques that will enable us to apply Minkowski's geometric point of view.

First, look again at the quadratic form (8.2). Do you realize that we can express (8.2) as a *linear combination of squares of two linear forms*? By simply adding and subtracting the term $(b^2/a)y^2$, then substituting the discriminant d, we obtain

$$\begin{aligned}
f(x,y) &= ax^2 + 2bxy + \frac{b^2}{a}y^2 + cy^2 - \frac{b^2}{a}y^2 \\
&= a\left(x^2 + \frac{2b}{a}xy + \frac{b^2}{a^2}y^2\right) + \left(\frac{ac - b^2}{a}\right)y^2 \\
&= a\left(x + \frac{b}{a}y\right)^2 + \frac{d}{a}y^2
\end{aligned}$$
(8.3)

In fact, it is possible to express $f(x,y)$ as a combination of squares of two linear forms in infinitely many ways. To see this, first subject $f(x,y)$ to an arbitrary linear transformation

$$\begin{aligned}
x &= \alpha x_1 + \beta y_1, \\
y &= \gamma x_1 + \delta y_1,
\end{aligned}$$
(8.4)

where $\alpha, \beta, \gamma, \delta$ are integers and $\Delta = \alpha\delta - \beta\gamma = \pm 1$. This linear transformation changes $f(x,y)$ into the form

$$g(x_1, y_1) = a_1 x_1^2 + 2b_1 x_1 y_1 + c_1 y_1^2,$$

where

$$\begin{aligned}
a_1 &= a\alpha^2 + 2b\alpha\gamma + c\gamma^2, \\
b_1 &= \left(a\alpha\beta + b(\alpha\delta + \beta\gamma) + c\gamma\delta\right), \quad \text{and} \\
c_1 &= a\beta^2 + 2b\beta\delta + c\delta^2.
\end{aligned}$$

Once again, we can complete the square to express $g(x_1, y_1)$ in the form (8.3).

We now claim that $f(x,y)$ and $g(x_1, y_1)$ are essentially the same quadratic form, in the sense that they represent exactly the same set of integers, n. In fact, suppose that $f(p,q) = n$ for some pair of integers (p,q); that is,

$$n = f(p,q) = ap^2 + 2bpq + cq^2.$$

Then, solving the system of linear equations (8.4) for x_1 and y_1, we see that

$$x_1 = \frac{1}{\Delta}(\delta x - \beta y) = (\pm 1)(\delta p - \beta q) = p_1 \quad \text{and}$$

$$y_1 = \frac{1}{\Delta}(-\gamma x + \alpha y) = (\pm 1)(-\gamma p + \alpha q) = q_1$$

are integers and, furthermore, that $g(p_1, q_1) = n$.

Remember that we can apply Minkowski's Fundamental Theorem to an M-set centered at the origin with area greater than or equal to 4 to obtain lattice points other than the origin contained inside or on the boundary. Hence, we want to formulate our algebraic problem—namely, finding the minimum positive value obtained by a quadratic form with positive discriminant—as a geometric one. That is, we wish to find an associated M-set that will fulfill the conditions of Minkowski's Fundamental Theorem.

To begin, note that, given the necessary and sufficient condition that $d > 0$, we know that all nonzero values $f(x, y) = n$ are positive integers, so they can be written as squares of real numbers. Thus,

$$f(x, y) = a \left(x + \frac{b}{a} y \right)^2 + \frac{d}{a} y^2 = s^2. \tag{8.5}$$

Because $f(-x, -y) = f(x, y) = s^2$, the ellipse described by (8.5) is symmetric about the origin; hence, it is an M-set. We now wish to find its area. Using the transformation

$$X = x + \frac{b}{a} y$$
$$Y = y, \tag{8.6}$$

(8.5) becomes

$$aX^2 + \left(\frac{d}{a} \right) Y^2 = s^2,$$

or

$$\frac{X^2}{\dfrac{s^2}{a}} + \frac{Y^2}{\dfrac{s^2 a}{d}} = 1.$$

This latter equation is an ellipse centered at the origin in the XY-plane with its major and minor axes coincident with the X- and Y-axes. In fact, it too is an M-set, and because the transformation (8.6) has determinant 1, the area of this ellipse is equal to the area of the rotated ellipse described by (8.5) in the xy-plane.

We will now prove the following lemma, which gives the formula for the area of an ellipse.

Lemma 8.1. *Suppose that a given ellipse, E, is described by the equation $(X^2/M^2) + (Y^2/N^2) = 1$. Then the area of E is given by the formula $A_E = \pi M N$.*

Proof. Consider the linear transformation given by $X_1 = (t/M)X$, $Y_1 = (t/N)Y$. Under this change of variables, the equation for the ellipse, E, is transformed into that for the circle

$$C: \quad X_1^2 + Y_1^2 = t^2,$$

with area $A_C = \pi t^2$. On the other hand, the determinant of the transformation is t^2/MN, so if we choose t at the outset to satisfy the condition $t^2 = MN$, the determinant is 1 and the areas of the ellipse and the circle are equal. This proves the assertion of the lemma.

We now see that, under the linear transformation above, the quadratic form $f(x, y) = s^2$ corresponds to the equation of an ellipse with area

$$A = \pi \left(\frac{s}{\sqrt{a}} \right) \left(\frac{s\sqrt{a}}{\sqrt{d}} \right) = \frac{\pi s^2}{\sqrt{d}}$$

and, hence, defines an M-set with that same area.

At this point, we invoke Minkowski's Fundamental Theorem. Suppose that we set the area of the ellipse (8.5) equal to 4; that is,

$$\frac{\pi s^2}{\sqrt{d}} = 4, \quad \text{or} \quad s^2 = \frac{4}{\pi}\sqrt{d}.$$

Then there exists at least one lattice point $(x, y) = (p, q) \neq (0, 0)$ inside or on $f(x, y) = s^2$ such that

$$ap^2 + 2bpq + cq^2 \leq s^2 = \frac{4}{\pi}\sqrt{d}.$$

The following theorem summarizes this result.

Theorem 8.1. *If $a > 0$ and $d = ac - b^2 > 0$, then there exists at least one pair of integers p, q, not both zero, such that*

$$f(p, q) = ap^2 + 2bpq + cq^2 \leq \frac{4}{\pi}\sqrt{d}.$$

8.3 An Improved Upper Bound

The constant $M_1 = (4/\pi)\sqrt{d}$ in Theorem 8.1 is not the best possible upper bound for our desired minimum value, M. Although Minkowski's geometrical method is powerful, it simply cannot find the smallest permissible value of M_1. To improve M_1, we will adopt an algebraic method developed by two prominent Russian mathematicians of the nineteenth century, Korkine and Zolotareff [**15**] For a less superficial survey of their method than we will attempt here, see [**13**, Ch. 2].

Let $(\mathbb{X}, \mathbb{Y}) = (p, q)$ be the lattice point giving the least numerical value of the positive definite quadratic form

$$f(\mathbb{X}, \mathbb{Y}) = A\mathbb{X}^2 + 2B\mathbb{X}\mathbb{Y} + C\mathbb{Y}^2, \tag{8.7}$$

where $A > 0$ and $d_1 = AC - B^2 > 0$. Clearly, p and q must be relatively prime, for if g.c.d.$(p, q) = s > 1$, then $p = sp_1$ and $q = sq_1$, in which case

$$f(p, q) = s^2 f(p_1, q_1).$$

That is, $f(p_1, q_1)$ has a smaller numerical value than $f(p, q)$.

If, however, g.c.d.$(p, q) = 1$, then two integers m and n exist such that $pm + qn = 1$. Hence the linear transformation

$$\mathbb{X} = px - ny,$$

$$\mathbb{Y} = qx + my$$

has determinant $\Delta = pm + qn = 1$, and it transforms the lattice point $(\mathbb{X}, \mathbb{Y}) = (p, q)$ into the lattice point $(x, y) = (1, 0)$. The quadratic form (8.7) is transformed into the quadratic form

$$g(x, y) = ax^2 + 2bxy + cy^2, \tag{8.8}$$

which, since $\Delta = 1$, takes on the same numerical values as $f(\mathbb{X}, \mathbb{Y})$.

We write (8.8) in the form

$$g(x, y) = a\left(x^2 + 2\frac{b}{a}xy + \frac{c}{a}y^2\right)$$
$$= a\left[(x + b'y)^2 + (c' - b'^2)y^2\right], \qquad \text{where } b' = \frac{b}{a}, c' = \frac{c}{a}.$$

Since $f(p, q)$ is minimal, $g(1, 0) = a$ is precisely that minimal value. Therefore, the numerical value of $g(x, y)$ is greater than or equal to $a > 0$ for all lattice points (x, y) other than $(0, 0)$.

Suppose $|b'| \geq \frac{1}{2}$ and let k be the nearest integer to b'. We will set $k - b' = \epsilon$; thus, $|k - b'| = |\epsilon| \leq \frac{1}{2}$. Now we perform another linear transformation,

$$x = x' - ky',$$

$$y = y',$$

with determinant $\Delta = 1$. This changes $g(x, y)$ into the form $Q(x', y')$, where

$$Q(x', y') = a\left[\left(x' - ky' + b'y'\right)^2 + \left(c' - b'^2\right)y'^2\right]$$
$$= a\left[\left(x' - \epsilon y'\right)^2 + \left(c' - b'^2\right)y'^2\right] \geq a,$$

for all lattice points (x', y').

Let's examine the particular lattice point $(x', y') = (0, 1)$. We see at once that

$$Q(0, 1) = a\left[\epsilon^2 + \left(c' - b'^2\right)\right] = a\epsilon^2 + a\left(c' - b'^2\right) \geq a$$

so that

$$a\left(c' - b'^2\right) \geq a - a\epsilon^2 = a\left(1 - \epsilon^2\right),$$

or

$$a\left(\frac{c}{a} - \frac{b^2}{a^2}\right) = \frac{ac - b^2}{a} = \frac{d}{a} \geq a\left(1 - \epsilon^2\right).$$

What is the value of the discriminant d? We have that

$$\epsilon = k - b' = k - \frac{b}{a} \quad \text{and} \quad \epsilon^2 = \left(k - \frac{b}{a}\right)^2 \leq \frac{1}{4}.$$

This gives us

$$a\left(1 - \epsilon^2\right) = a\left[1 - \left(k - \frac{b}{a}\right)^2\right] \geq \frac{3}{4}a.$$

Consequently, the value of the discriminant is

$$d \geq \frac{3}{4}a^2, \quad \text{or} \quad a \leq \frac{2}{\sqrt{3}}\sqrt{d}.$$

The absolute minimum of $g(x, y)$ being a, we have

$$a = M_2\sqrt{d}, \quad \text{or} \quad M_2 = \frac{a}{\sqrt{d}} \leq \frac{2}{\sqrt{3}}.$$

But recall that the minimum value of the special form $x^2 + xy + y^2$ is 1, that $b = \frac{1}{2}$, and that its discriminant $d = \frac{3}{4}$. Then $a = 1 \leq M_2\sqrt{\frac{3}{4}}$ cannot hold unless $M_2 \geq \sqrt{\frac{4}{3}}$. Thus, finally, we must have $M_2 = 2/\sqrt{3}$.

We will reach the same conclusion even if we change our assumptions, letting $|b'| < \frac{1}{2}$ and $k = 0$ in the above discussion. Hence we have proved the next theorem, which refines Theorem 8.1.

Theorem 8.2. *Let* $f(x, y) = ax^2 + 2bxy + cy^2$, *with* $a > 0$ *and* $d = ac - b^2 > 0$, *be a positive definite quadratic form. Then integers* p, q, *not both zero, exist such that*

$$|f(p, q)| = f(p, q) \le \frac{2}{\sqrt{3}} \sqrt{d}.$$

Example. The quadratic form

$$f(x, y) = 25x^2 + 126xy + 162y^2$$

must have a minimum value

$$\le \frac{2}{\sqrt{3}} \sqrt{(25)(162) - (63)^2} = 10.39\ldots.$$

Since the coefficients in $f(x, y)$ are integers, the minimum is less than or equal to 10.

8.4 (Optional) Bounds for the Minima of Quadratic Forms in More Than Two Variables

The following remarks are included for those readers who are familiar with the vocabulary of quadratic forms. All results are stated without proof, but references are cited for the interested reader to pursue. The average reader will lose little by ignoring this section.

Section 8.3 gave an introduction to the methods developed by Korkine and Zolotareff for bounding the minimal nonzero values of a positive definite quadratic form in two variables, otherwise known as a positive definite *binary* quadratic form. The two also invented an ingenious method for discussing quadratic forms in more than two variables; they described their procedure as "developing a quadratic form according to its minima" [**15**].

Let X represent an n-tuple; that is, $X = (x_1, x_2, \ldots, x_n)$. Let $Q(X) = Q(x_1, x_2, \ldots, x_n)$ be a positive definite quadratic form with discriminant $d > 0$. We wish to answer the following question, which generalizes that motivating the work of Sections 8.2 and 8.3:

> Let $\epsilon > 0$ *be arbitrary. What is the smallest positive real number* M_n *such that there exists a nonzero n-tuple* $X \in \mathbb{Z}^n$ *for which* $Q(X) \le M_n + \epsilon$?

If this lower bound is actually achieved by a form $Q(X)$ with discriminant d, we speak of the *minimum* of $Q(X)$ for $X \ne (0)$. Because this value takes the form of a constant times a fractional power of d, we will denote that constant by m_n.

Minkowski [**18**] proved via his geometrical methods that if $Q(X)$ is a positive definite quadratic form in n variables, then there exist n-tuples of integers X such that

$$0 < Q(X) \le \frac{4}{\pi} \left[\Gamma \left(1 + \frac{n}{2} \right) \right]^{2/n} d^{1/n} = m_n d^{1/n} = M_n,$$

where $\Gamma(x)$ denotes the ordinary gamma function.

In 1914, H. F. Blichfeldt [**1**] (of whom we shall hear more in Chapter 9), using a principle he had newly developed in the geometry of numbers, was able to replace Minkowski's result by

$$0 < Q(X) \le \frac{2}{\pi} \left[\Gamma \left(1 + \frac{n+2}{2} \right) \right]^{2/n} d^{1/n} = M_n' d^{1/n}.$$

How big an improvement is this? In Minkowski's result, the asymptotic value of m_n is $(2n)/(\pi e)$; that is, $m_n \to (2n)/(\pi e)$ as $n \to \infty$. Symbolically, we write $m_n \sim (2n)/(\pi e)$. But in Blichfeldt's result, $m_n' \sim n/(\pi e)$—which is one-half that of Minkowski's limit! Minkowski also proved that the asymptotic value of m_n cannot fall below $n/(2\pi e)$.

The minimum value m_2 for $n = 2$ was first determined by Hermite [**9**] and was known to Gauss [**7**]. The minima for $n = 3, 4, 5$ were found by Korkine and Zolotareff [**15, 16, 17**], and Blichfeldt [**2**] determined the minima for $n = 6, 7, 8$. These minimum values are:

$$m_2 = \frac{2}{\sqrt{3}}, \quad m_3 = \sqrt[3]{2}, \quad m_4 = \sqrt{2}, \quad m_5 = \sqrt[5]{8},$$

$$m_6 = \sqrt[6]{\frac{64}{3}}, \quad m_7 = \sqrt[7]{64}, \quad m_8 = 2.$$

8.5 Approximating by Rational Numbers

Theorem 8.2 can be applied to the problem of approximating *irrational numbers by rational numbers*. In fact, it is easy to prove that there exist infinitely many rational fractions m/n, for $n \ne 0$, such that

$$\left| \alpha - \frac{m}{n} \right| \le \frac{2}{\sqrt{3}n^2},$$

where α is any real number. The degree of approximation is proportional to $1/n^2$, so we get rather good approximations. For details, see [**13**, p. 40].

Referring to expression (8.5) for the quadratic form $f(x,y)$, we can take it an additional step, writing

$$f(x,y) = \left(\sqrt{a}x + \frac{b}{\sqrt{a}}y\right)^2 + \left(\sqrt{\frac{d}{a}}y\right)^2.$$

Note that, in this form, the coefficients of the polynomials in the parentheses are no longer rational. However, once we make this allowance, we see that we can write every quadratic form as a sum of squares of linear expressions. Using this, we let α be any irrational number and consider the quadratic form

$$Q(m,n) = \left(\frac{\alpha n - m}{\epsilon}\right)^2 + \epsilon^2 n^2$$
$$= \frac{1}{\epsilon^2}m^2 - 2\frac{\alpha}{\epsilon^2}mn + \left(\frac{\alpha}{\epsilon^2} + \epsilon^2\right)n^2,$$

where m and n are integers, and whose discriminant is

$$d = \frac{1}{\epsilon^2}\left(\frac{\alpha}{\epsilon^2} + \epsilon^2\right) - \frac{\alpha^2}{\epsilon^4} = 1.$$

Here ϵ is an arbitrary positive number.

By Theorem 8.2, we can always find two integers m and n, not both zero, such that

$$\left(\frac{\alpha n - m}{\epsilon}\right)^2 + \epsilon^2 n^2 \leq \frac{2}{\sqrt{3}}.$$

These two inequalities follow:

$$\left|\alpha - \frac{m}{n}\right| \leq \frac{\epsilon}{|n|}\sqrt{\frac{2}{\sqrt{3}}},$$
$$\qquad\qquad\qquad \text{where } n > 0. \qquad (8.9)$$
$$|n| \leq \frac{1}{\epsilon}\sqrt{\frac{2}{\sqrt{3}}},$$

Since α is irrational, we cannot have $|\alpha - (m/n)| = 0$. Hence there must exist infinitely many rational fractions m/n, where $n > 0$, satisfying the first inequality in (8.9). To see this, simply assign smaller and smaller values to ϵ; for each value of ϵ a corresponding rational fraction m/n exists. These fractions cannot all be equal, for $|\alpha - (m/n)| \to 0$ as $\epsilon \to 0$.

If we use the second inequality in (8.9) to eliminate ϵ in the first, we obtain

$$\left|\alpha - \frac{m}{n}\right| \leq \frac{2}{\sqrt{3}n^2}, \qquad \text{where } n > 0. \tag{8.10}$$

Hence we have proved that there exist infinitely many rational fractions m/n approximating α, the accuracy of approximation being inversely proportional to the square of the denominator n.

Of course, this is far from a "best possible" result. Hurwitz [**14**] proved that for any irrational number α, there exist infinitely many rational fractions m/n, for $n > 0$, such that

$$\left|\alpha - \frac{m}{n}\right| < \frac{1}{\sqrt{5}n^2}.$$

Here, $\sqrt{5}$ is the best possible constant, in the sense that the theorem becomes false if any larger number is substituted for $\sqrt{5}$. Comparing the constants, we observe that $2/\sqrt{3} = 1.14155\ldots$, while $1/\sqrt{5} = 0.4472\ldots$, which is smaller. Still, we can obtain rather good approximations from (8.10) even with comparatively small denominators.

8.6 Sums of Four Squares

In 1621 Claude Bachet de Méziriac (1581–1638), an enthusiastic popularizer of Diophantus, stated without proof that every positive integer n is the sum of four integral squares. That is, every positive integer n can be expressed in the form

$$n = x^2 + y^2 + z^2 + w^2.$$

where x, y, z, w are integers.

Bachet's assertion is not an easy theorem to prove. Fermat left one of his now-famous marginal notes stating that he had a proof; while this is probably true, we will never know for sure. The great and resourceful Euler tried repeatedly between 1730 and 1750 to supply a proof, only to fail. When at last Lagrange published the first proof in 1770, he acknowledged his great indebtedness to the pathbreaking labors of Euler, whom he had succeeded at Frederick the Great's Prussian Academy in Berlin [**5**].

We will state and prove Lagrange's Theorem.

Theorem 8.3 (Lagrange's Theorem). *Every positive integer is representable as the sum of four integral squares.*

Many a mathematician has attempted a proof of Lagrange's Theorem. The following proof, due to Davenport [3], is somewhat similar to that of Grace [8] and is a modification of one given by Hermite in 1853 [10, 11]; see also [12]. Unlike Davenport, Hermite could not appeal to Minkowski's Fundamental Theorem (which did not yet exist); instead he used his own results on the minimum of positive definite quadratic forms.

Davenport did not claim that this is the ideal proof of Lagrange's Theorem, but it serves our purposes well. Not only is it fairly simple, with a minimum of details, but it is an excellent example of how the geometry of numbers can be used to prove a purely arithmetical result.

All direct elementary proofs of Lagrange's theorem require the following lemma.

Lemma 8.2. *For any odd positive integer m, integers a and b exist such that $a^2 + b^2 + 1 = mk$, where k is some integer.*

In short, $a^2 + b^2 + 1$ is divisible by m. Those familiar with congruence notation would write $a^2 + b^2 + 1 \equiv 0 \pmod{m}$.

We will take this lemma for granted, since its proof would draw us into deep waters. It involves "quadratic residues" when m is a prime p, an inductive argument on ν when $m = p^\nu$, and a combination of these results for general m [4].

Proof of Lagrange's Theorem. We begin by defining four linear forms $\mathbb{X}, \mathbb{Y}, \mathbb{Z}, \mathbb{W}$ in the four variables x, y, z, w:

$$\mathbb{X} = mx \qquad + az + bw,$$

$$\mathbb{Y} = \qquad my + bz - aw,$$

$$\mathbb{Z} = \qquad\qquad z,$$

$$\mathbb{W} = \qquad\qquad\qquad w.$$

The determinant of these linear forms is

$$\Delta = \begin{vmatrix} m & 0 & a & b \\ 0 & m & b & -a \\ 0 & 0 & 1 & 0 \\ 0 & 0 & 0 & 1 \end{vmatrix} = m^2. \qquad (8.11)$$

It is easy to see that $\Delta = m^2$, once we notice that this is a "triangular" determinant, with zeros below the principal diagonal. Thus its value is simply the product of the numbers $m, m, 1, 1$ on this diagonal.

Now let x, y, z, w take on all integral values $0, \pm 1, \pm 2, \ldots$. The corresponding points $(\mathbb{X}, \mathbb{Y}, \mathbb{Z}, \mathbb{W})$ will form a lattice in four-dimensional space of determinant $\Delta = m^2$.

We want to calculate the sum of the squares $\mathbb{X}^2 + \mathbb{Y}^2 + \mathbb{Z}^2 + \mathbb{W}^2$. Applying the equations (8.11) in a simple (but tedious) calculation, we find that $\mathbb{X}^2 + \mathbb{Y}^2 + \mathbb{Z}^2 + \mathbb{W}^2$ is equal to

$$m(mx^2 + my^2 - 2axy - 2ayw + 2bxy + 2byz) +$$
$$(a^2 + b^2 + 1)w^2 + (a^2 + b^2 + 1)y^2.$$

The first of these terms is divisible by m and, by virtue of Lemma 8.2, the next two terms are also divisible by m. Hence for all values of x, y, z, w we have

$$\mathbb{X}^2 + \mathbb{Y}^2 + \mathbb{Z}^2 + \mathbb{W}^2 = km, \tag{8.12}$$

where k is some positive integer.

Suppose we could prove that there exists a lattice point other than 0 for which

$$\mathbb{X}^2 + \mathbb{Y}^2 + \mathbb{Z}^2 + \mathbb{W}^2 < 2m. \tag{8.13}$$

Then it would follow from (8.12) that integers, not all zero, must exist such that

$$\mathbb{X}^2 + \mathbb{Y}^2 + \mathbb{Z}^2 + \mathbb{W}^2 = m.$$

This would prove Lagrange's theorem for m odd, and this is what we will now do, taking advantage of Minkowski's geometric point of view.

The inequality (8.13) represents a sphere in 4-space of radius $\sqrt{2m}$. It can be shown by integration that the volume of a four-dimensional sphere of radius r is

$$\frac{1}{2}\pi^2 r^4 = \frac{1}{2}\pi^2 (2m)^2.$$

By the general form of Minkowski's Fundamental Theorem in n-dimensions (Theorem 5.3), it will suffice to show that the volume of this sphere is greater than $2^4\Delta = 2^4 m^2$. Symbolically, we must show that

$$\frac{1}{2}\pi^2 (2m)^2 > 2^4 m^2,$$

which simplifies to $\pi^2 > 8$, a true inequality.

This proves Lagrange's Theorem for any odd positive integer m. The result can be extended at once to even integers, for if

$$m = \mathbb{X}^2 + \mathbb{Y}^2 + \mathbb{Z}^2 + \mathbb{W}^2,$$

then

$$2m = (\mathbb{X} + \mathbb{Y})^2 + (\mathbb{X} - \mathbb{Y})^2 + (\mathbb{Z} + \mathbb{W})^2 + (\mathbb{Z} - \mathbb{W})^2.$$

We end this discussion by stating without proof another theorem.

Theorem 8.4. *The total number of representations of a positive integer n as the sum of four squares, representations that differ only in order and sign being counted as distinct, is eight times the sum of the divisors of n that are not multiples of* 4.

Mimicking the notation of Section 4.3, we write

$$R_4(n) = R(n = p^2 + q^2 + r^2 + s^2),$$

the number of distinct representations of n as a sum of four squares. Then Theorem 8.4 is an extension of Theorem 4.4 and, symbolically, we write

$$R_4(n) = 8 \sum_{\substack{d|n \\ 4 \nmid d}} d.$$

To illustrate, note that

$$6 = (\pm 2)^2 + (\pm 1)^2 + (\pm 1)^2 + 0^2.$$

For that ordering of squares, we have $2^3 = 8$ possible choices of signs. Furthermore, there are $\binom{4}{2} = 6$ possible positions for the pair of ones, followed by two possible remaining positions for the zero. Hence, 6 has $8 \cdot 6 \cdot 2$ representations as a sum of four squares. On the other hand, the divisors d of 6 are $d = 1, 2, 3, 6$, and we see that

$$R_4(6) = 8 \sum_{\substack{d|6 \\ d \nmid 4}} d = 8(1 + 2 + 3 + 6) = 96.$$

References

1. H. F. Blichfeldt, "A New Principle in the Geometry of Numbers with Some Applications," *Transactions of the AMS* 15:3 (July 1914):227–35.
2. _____, "The Minimum Values of Positive Quadratic Forms in Six, Seven, and Eight Variables," *Mathematische Zeitschrift* 39 (1934):1–15.
3. Harold Davenport, "The Geometry of Numbers," *Math. Gazette* 31 (1947):206–10.
4. _____, *The Higher Arithmetic* (New York: Dover, 1983), 124.
5. L. E. Dickson, Preface to *History of the Theory of Numbers, Vol. II: Diophantine Analysis* (Washington, D.C.: Carnegie Institute, 1920), x.
6. _____, *History of the Theory of Numbers, Vol. III: Quadratic and Higher Forms* (Washington, D.C.: Carnegie Institute, 1923).
7. C. F. Gauss, *Werke* (Göttingen: Gesellschaft der Wissenschaften, 1863–1933).
8. J. H. Grace, "The Four Square Theorem," *Journal of the London Mathematical Society* 2 (1927):3–8.
9. Charles Hermite, "Lettres de Hermite à M. Jacobi," *J. reine angew. Math.* 40 (1850):261–315.
10. _____, *Comptes Rendus Paris* 37 (1853).
11. _____, *J. reine angew. Math.* 47 (1854):343–5, 364–8.
12. _____, *Oeuvres*, Vol. I (Paris: E. Picard, 1905), 288.
13. David Hilbert and S. Cohn-Vossen, *Geometry and the Imagination*, translated by P. Nemenyi (New York: Chelsea, 1952).
14. A. Hurwitz, "Über die angenäherte Darstellung der Irrationalzahlen durch rationale Brüche," *Mathematische Annalen* 39 (1891):279–84.
15. A. Korkine and E. I. Zolotareff, "Sur les formes quadratiques positives quaternaires," *Mathematische Annalen* 5 (1872):581–3.
16. _____, "Sur les formes quadratiques," *Mathematische Annalen* 6 (1873):366–89.
17. _____, "Sur les formes quadratiques positives," *Mathematische Annalen* 11 (1877):242–92.
18. Hermann Minkowski, "Über die positiven quadratischen Formen un über kettenbruchähnliche Algorithm," *J. reine agnew. Math.* 107 (1891):209–12.

9
A New Principle
in the Geometry of Numbers

9.1   Blichfeldt's Theorem

Around 1891, Hermann Minkowski discovered his Fundamental Theorem, opening up a new field of study which he called the *geometry of numbers*. Using his theorem and its generalizations, Minkowski was able to solve many difficult problems in number theory. In Chapter 6 we examined some of the easier applications of Minkowski's theorems.

Despite the excitement aroused by Minkowski's groundbreaking work, it was another 15 years before any new principle in the geometry of numbers was discovered. The credit for this breakthrough goes to Hans Frederik Blichfeldt, who in 1914 published a theorem from which a great portion of the geometry of numbers follows. Once stated, this theorem seems almost intuitively obvious, yet it is so powerful that it carried Blichfeldt, and others after him, to proofs of results not attainable using Minkowski's original theorems. The reader will find a short biography of Blichfeldt, along with that of Minkowski, in Appendix III.

This chapter is an introduction to Blichfeldt's Theorem. For simplicity, we shall formulate it in terms of two-dimensional space, $\mathbb{R}^2$, rather than Blichfeldt's generalized n-dimensional space, $\mathbb{R}^n$. After proving Blichfeldt's Theorem, we will look at a few things it lets us do with Minkowski M-sets.

Some concepts must first be defined. The process called *translating* means relocating a planar figure to a new set of axes with a different origin, so that its lattice points have new coordinates. The relocated structure preserves the original size and shape, but it can then be expanded or contracted as desired. *Rotating* the axes about the origin means turning them to a new alignment while preserving their angles. *Parallel displacement* is simply translation without any rotation of the axes; that is, the original orientation is preserved. Finally, as in Section 3.3, a point set *covers* a lattice point if the set contains the lattice point within it or on its boundaries.

Theorem 9.1 (Blichfeldt's Theorem). *If the area A of a two-dimensional set C is greater than the integer n, then, by a parallel displacement, C can be made to cover at least $n + 1$ lattice points of Λ.*

It follows at once from the theorem that if C has an area $A = n$, then, by a translation, C can be made to cover n lattice points.

9.2   Proof of Blichfeldt's Theorem

We will prove Blichfeldt's Theorem by a geometrical analysis in which we slice, stack, and puncture a planar region—figuratively, of course.

Let C be any planar region or set of points, placed anywhere on the point lattice Λ. We need not assume that C is convex, nor that it has central symmetry. The squares of the lattice Λ divide C into parts $C_1, C_2, \ldots, C_k$, each lying in its corresponding square $R_1, R_2, \ldots, R_k$. Figure 9.1a depicts a simplified set C, where the numbers 1 through 5 refer to the sets $C_1, \ldots, C_5$.

Now we select any other faraway square R of Λ, and by parallel displacements we translate each of the squares R_i onto it. The adjacent parts $C_1, C_2, \ldots, C_k$ of these squares are transformed into congruent parts of R as suggested in Figure 9.1b. We can think of these squares as being piled layer by layer upon the square R.

The following lemma is necessary before we can proceed.

Lemma 9.1. *The square R contains at least one point X that is covered by the parts $C_1, C_2, \ldots, C_k$ no fewer than $n + 1$ times.*

Proof of Lemma 9.1. Suppose that each point of the square R were covered no more than n times by the parts $C_1, C_2, \ldots, C_k$. Then the combined areas of these parts could not exceed n. Consequently, the combined areas of the parts $C_1, C_2, \ldots, C_k$, when translated back to their original positions in C would not exceed n. But this contradicts our assumption that C has an area $A > n$. Hence R must have at least one point X that is covered no fewer than $n+1$ times by the sets C_i, which proves the lemma.

For the next step in our proof of Blichfeldt's Theorem, imagine poking a needle straight down through these $n + 1$ sets (or layers) on R. Each

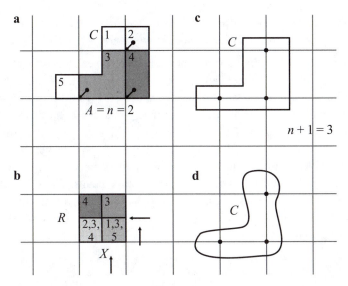

Figure 9.1
Proving Blichfeldt's Theorem.

set will now have a small hole in it, and all the holes must have the same relative position with respect to the square R. What will happen when we translate all the squares in R back to their original places in C? Clearly, the pinholes still keep the same relative position with respect to their individual squares as when they were in R. Hence, any translation that carries any pinhole to cover a lattice point must also carry the other pinholes to cover lattice points.

Examine the "pinholes" translated back from Figure 9.1b to Figure 9.1a. Sliding C slightly down and left gives us Figure 9.1c, in which the pinholes cover lattice points. Since we have at least $n + 1$ pinholes in C, a translation must move C so that it covers at least $n + 1$ lattice points. This proves Theorem 9.1.

In this illustration, we used rectangles to make things easy to see. Of course, the same proof holds for any set C, such as that in Figure 9.1d.

9.3   A Generalization of Blichfeldt's Theorem

We state without proof a more general theorem.

Theorem 9.2.   *A bounded set of points C, with area A, can be translated on* Λ *so as to cover a number of lattice points greater than A.*

Theorem 9.2 is the best possible, except for special C-sets. Do you see why? Consider the rectangle in Figure 9.2, with sides parallel to the x- and y-axes and of dimensions $n - \epsilon_1$ and $1 - \epsilon_2$. Here, n is a positive integer, while ϵ_1 and ϵ_2 are positive numbers that can be made as small as we please. The area of C is $(n - \epsilon_1)(1 - \epsilon_2) = n - \epsilon$, where $\epsilon \to 0$ as $\epsilon_1 \to 0$ and $\epsilon_2 \to 0$. Hence, $n - 1 < A < n$. By Theorem 9.1, we can translate C until it contains n or more lattice points. However, as the figure shows, it cannot contain more than n lattice points. This proves that Theorem 9.2 is best possible, in the sense that we have constructed a set C that covers no more than $[A] + 1$ lattice points. Thus, that lower bound cannot be increased.

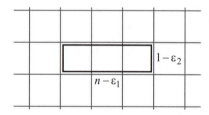

Figure 9.2
A set C with area less than, but arbitrarily close to, n.

Both Theorems 9.1 and 9.2 follow as special cases from Blichfeldt's general theorem of 1914. Not only did Blichfeldt deal with n-dimensional space, $\mathbb{R}^n$, but he used a much more general definition of lattice points in $\mathbb{R}^n$. Today, Blichfeldt's Theorem is often stated in the following form, using the concept of *difference points*.

Theorem 9.3.   *Let C be a bounded set of points in* $\mathbb{R}^2$ *whose area exceeds* 1. *Then there exist at least two points* $P_1 : (x_1, y_1)$ *and* $P_2 : (x_2, y_2)$ *in C such that the "difference point"* $P : (x_1 - x_2, y_1 - y_2)$ *has integral coordinates.*

Proof.   Place C anywhere on Λ. Blichfeldt's Theorem tells us that we can translate C to a new position C' so that it covers at least two lattice points, say P_1' and P_2'. See Figure 9.3. Observe that the lengths $|\overline{P_2'P'}|$ and $|\overline{P'P_1'}|$

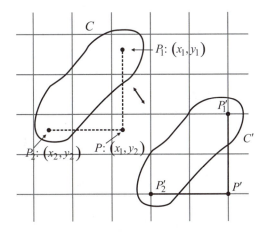

Figure 9.3
Translating C to C' and back.

are integers and are equal to the lengths $|\overline{P_2P}|$ and $|\overline{PP_1}|$. Thus, when C' is translated back to C, $x_2 - x_1$ and $y_2 - y_1$ are also both integers.

9.4   A Return to Minkowski's Theorem

Suppose the set C in Theorem 9.1 is a Minkowski M-set—that is, a convex point set symmetric with respect to point O, its center of symmetry. Then Minkowski's Fundamental Theorem follows from Blichfeldt's Theorem. To see how, let's state Minkowski's Theorem again and prove it using Blichfeldt's result.

Theorem 9.4 (Minkowski).   *If M is a bounded convex symmetrical set with an area greater than 4, then it contains at least one lattice point other than the origin.*

Proof.   Starting with a Minkowski M-set centered at O, we shrink it, preserving similarity, until its area is equal to $1 + \epsilon$, where ϵ is a positive number to be specified later. By Theorem 9.1, this set can be translated to position M' so that it covers two (or more) lattice points P_1 and P_2. See Figure 9.4.

Let O' be the center of symmetry of M' after this translation, and let P_1' and P_2' be the lattice points diametrically opposite to P_1 and P_2. By the

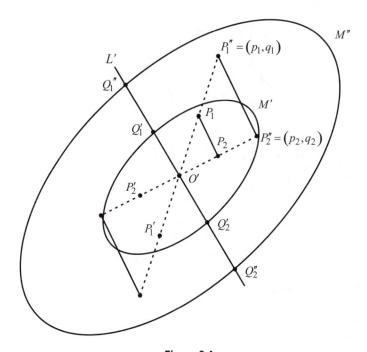

Figure 9.4

Contraction and translation of M to M', expansion to M''.

convexity of M', the parallelogram $P_1 P_2 P'_1 P'_2$ will be entirely contained in the point set defined by M' and its boundary.

Now draw a line L' through O' and parallel to $P_1 P_2$, so that it cuts the boundary of M' in points Q_1 and Q_2. Because P_1 and P_2 are lattice points, their coordinates (p_1, q_1) and (p_2, q_2) are integral. Hence, L' is a line with rational slope $(q_2 - q_1)/(p_2 - p_1)$. Clearly, by the convexity of M', the quadrilateral $Q_1 Q_2 P_2 P_1$ is contained in M, and so the segment $\overline{Q_1 Q_2}$ has length greater than or equal to the length of $\overline{P_1 P_2}$; that is,

$$|\overline{Q_1 Q_2}| \geq |\overline{P_1 P_2}| = \sqrt{(p_2 - p_1)^2 + (q_2 - q_1)^2}.$$

Next we expand the dimensions of M', again preserving similarity, until its linear dimensions are doubled. This gives us the set M'' with an area of $4(1 + \epsilon) = 4 + 4\epsilon$. The line segments $\overline{Q_1 Q_2}$ and $\overline{P_1 P_2}$ are also doubled in length to form expanded segments $\overline{Q''_1 Q''_2}$ and $\overline{P''_1 P''_2}$, where Q''_1 and Q''_2 lie on the boundary of M'', while P''_1 and P''_2 lie inside or on the boundary of M''. Furthermore, since M has area greater than 4 and is

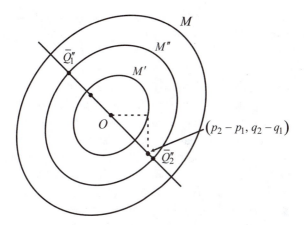

Figure 9.5
The retranslated M-sets.

similar to M'', we can chose ϵ in such a way that the translation of M to center O' contains M''.

We now translate M'' back to its original position, so that its center of symmetry O' is returned to the origin O and the line L' is translated into the line

$$L:\ y = \frac{q_2 - q_1}{p_2 - p_1}\ x$$

through the origin. We claim that at least two lattice points lie on the translated line segment $\overline{Q_1'' Q_2''}$. In fact, we see that the lattice points nearest to the origin on the line L lie at the points $\pm(p_2 - p_1, q_2 - q_1)$. But each of these points lies at distance

$$\sqrt{(p_2 - p_1)^2 + (q_2 - q_1)^2} \leq \left|\overline{Q_1 Q_2}\right| = \left|\overline{OQ_1}\right| = \left|\overline{OQ_2}\right|$$

from the origin. Hence both of these lattice points lie in M''.

Now, to be more specific than earlier, our original set M has area greater than 4, say $A_M = 4 + \delta$. If we chose $\epsilon < \delta/4$, then $A_{M''} < A_M$; hence, $M'' \subset M$, proving the statement of the theorem. See Figure 9.5.

9.5 Applications of Blichfeldt's Theorem

Blichfeldt's use of translation sheds light on many of Minkowski's Theorems, as the following example shows.

Theorem 9.5. *Suppose a Minkowski M-set centered at the origin O has area 4A. Then the M-set contains more than* $[A] - 1$ *pairs of lattice points other than the origin O.*

Proof. Suppose that we are given a Minkowski M-set of area A. Let $n - 1 < A \leq n$. Then by a translation the M-set can be made to cover $n > A$ lattice points. Translate it once more, so that one of these lattice points, say P, lands on the boundary of M. Let the center of this translated set be O', and extend the line $O'P$ to O'', making $O'P = PO''$. Now we construct an M-set, M', that is equal to M and similarly placed, but has O'' as its center of symmetry. See Figure 9.6a.

Clearly, P lies on the boundary of M'. Moreover, to every lattice point Q' of M, there is a corresponding lattice point Q'' in M'' where $Q'P = PQ''$. In general, Q'' will not also belong to M unless the line $Q'PQ''$ is part of a common boundary of both M and M'.

Ruling out this case for the moment, we count $2(n-1) + 1 = 2n - 1$ lattice points covered by M and M'. Now we translate M, so that O' coincides with P—but then we enlarge it, so that its dimensions are twice those of M. This new set, M'', has area $4A$ and it covers the $2n - 1$ lattice

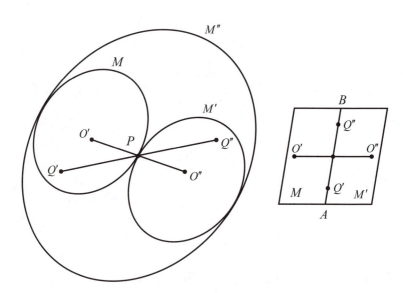

Figure 9.6
Corresponding lattice points in translations.

points mentioned above, including P. Hence, if $n > A$, then $n - 1$ pairs of lattice points, in addition to P, are covered by M''. The theorem is proved by simply translating M'' back to O.

What about the case when M' and M have a common boundary AB? See Figure 9.6b. Along this common line, let Q' be the lattice point nearest to an extremity of AB; say, nearest to A. Then, we can simply take Q' instead of P in the above argument, and the same conclusion follows.

We stress that Blichfeldt's theorem of 1914 is far more than just a steppingstone to Minkowski's result. Indeed, Blichfeldt's own applications of his theorem are many; however, they are also very difficult reading. We have already referred to one important application in optional Section 8.4, and we restate that application now in the following form.

Theorem 9.6. *Let f be a positive definite quadratic form of determinant $d > 0$. Then there exist integers $l_1, l_2, \ldots, l_n$, not all zero, such that $f(l_1, l_2, \ldots, l_n)$ is less than or equal to*

$$\frac{2}{\pi} \left[\Gamma \left(1 + \frac{n+2}{2} \right) \right]^{2/n} d^{1/n}.$$

Here Γ is the usual gamma function.

Blichfeldt conducted a long series of researches on the minima of quadratic forms, eventually leading to his famous publication, "The Minimum Values of Positive Quadratic Forms in Six, Seven, and Eight Variables" (*Math. Zeit.*, Vol. 39, 1934, pp. 1–15). His work represents a monumental achievement in its own right—a high point from which many new paths besides Minkowski's could be surveyed.

10
A Minkowski Theorem
(Optional)

10.1 A Brief History of the Question

In 1866, Pafnuty Livovich Tchebychev (1821–1894), one of Russia's greatest mathematicians, proved the following theorem.

Theorem 10.1. *Let θ be irrational and suppose that α is any real number for which the equation $x - \theta y - \alpha = 0$ has no solutions in integers p, q. Then for any given positive ϵ, there are infinitely many pairs of integers p, q such that*

$$\left| q(p - \theta q - \alpha) \right| < 2, \tag{10.1}$$

and, at the same time, such that $|p - \theta q - \alpha| < \epsilon$.

Tchebychev's proof was long, running to forty pages [7], and was purely arithmetical, making extensive use of continued fractions. Hermite [3] in 1879 proved that the constant 2 in (10.1) may be replaced by $\frac{1}{2}$ and, in fact, by $\sqrt{2/27}$. He also used purely arithmetical methods. The situation was not improved again until 1901, when Minkowski used his new geometrical methods to prove the following theorem.

Theorem 10.2. *If θ is irrational, and if α is any non-integer real number such that $x - \theta y - \alpha = 0$ has no solutions in integers, then, for any given positive number ϵ, there are infinitely many pairs of integers p, q, such that*

$$\left| q(p - \theta q - \alpha) \right| < \frac{1}{4},$$

and, at the same time, such that $|p - \theta q - \alpha| < \epsilon$.

10.2   A Proof of Minkowski's Theorem

The proof of Theorem 10.2 presented below is due to Blichfeldt, given in a set of lecture notes in 1936. Its principle device is to replace the nonhomogeneous linear form $x - \theta y - \alpha$ by the homogeneous form $x - \theta y - \alpha z$, obtained by introducing a third variable z, and then later to fix conditions so that $z = 1$. A second device is to introduce a positive parameter t.

On the ordinary x, y, z-coordinate system, consider the prism κ_t that is bounded by the six planes:

$$\kappa_t \; : \quad \begin{aligned} |x - \theta y - \alpha z| + t|y| &= \sqrt{t}, \\ |z| &= 2, \end{aligned} \tag{10.2}$$

with t a positive parameter at our disposal. At the outset, we take t such that

$$\sqrt{t} < \min\big\{\epsilon, \alpha - [\alpha], 1 - \alpha + [\alpha]\big\},$$

where again $[\alpha]$ is the largest integer $\le \alpha$. Thus, $\sqrt{t} < \frac{1}{2}$. Clearly, the prism κ_t is a Minkowski M-set in 3-space.

The six planes in (10.2) can be arranged into three pairs of parallel planes:

$$\begin{cases} x + (t - \theta)y - \alpha z + \sqrt{t} = 0, \\ x + (t - \theta)y - \alpha z - \sqrt{t} = 0, \end{cases}$$

$$\begin{cases} x + (-t - \theta)y - \alpha z + \sqrt{t} = 0, \\ x + (-t - \theta)y - \alpha z - \sqrt{t} = 0, \end{cases} \tag{10.3}$$

$$\begin{cases} z + 2 = 0, \\ z - 2 = 0. \end{cases}$$

Solid analytic geometry gives us a formula for the volume of a prism bounded by six planes, parallel in pairs. Let the equations of these planes be

$$\begin{cases} a_1 x + b_1 y + c_1 z + d_1 = 0, \\ a_1 x + b_1 y + c_1 z + d_2 = 0, \end{cases}$$

$$\begin{cases} a_2 x + b_2 y + c_2 z + e_1 = 0, \\ a_2 x + b_2 y + c_2 z + e_2 = 0, \end{cases}$$

$$\begin{cases} a_3 x + b_3 y + c_3 z + f_1 = 0, \\ a_3 x + b_3 y + c_3 z + f_2 = 0. \end{cases}$$

Then the volume of this prism is $|\nabla|$, where

$$\nabla = -\frac{(d_1 - d_2)(e_1 - e_2)(f_1 - f_2)}{\begin{vmatrix} a_1 & b_1 & c_1 \\ a_2 & b_2 & c_2 \\ a_3 & b_3 & c_3 \end{vmatrix}}. \tag{10.4}$$

(This derivation is best arrived at using vector analysis.) Applying formula (10.4) to the six planes in (10.3), we obtain

$$\nabla_t = -\frac{\left(2\sqrt{t}\right)\left(2\sqrt{t}\right)(4)}{\begin{vmatrix} 1 & t-\theta & -\alpha \\ 1 & -t-\theta & -\alpha \\ 0 & 0 & 1 \end{vmatrix}} = \frac{-16t}{\begin{vmatrix} 1 & t-\theta \\ 1 & -t-\theta \end{vmatrix}} = \frac{-16t}{-2t} = 8. \tag{10.5}$$

Note that the determinant in the denominator of (10.5) was expanded by the elements of its third row.

Since $\nabla_t = 2^3$, Minkowski's Fundamental Theorem tells us there must be at least one pair of nonzero lattice points $P_1 : (p_1, q_1, r_1)$ and $P_{-1} : (-p_1, -q_1, -r_1)$ inside or on the surface of κ_t. Clearly, since κ_t is bounded above and below by the planes $z = 2$ and $z = -2$, the only integer values that r_1 can have are $r_1 = \pm 2, \pm 1, 0$.

It is important, at this stage, to show that we can adjust things so that $|r_1| = 0$ or 1. To this end, suppose we shrink the dimensions of κ, uniformly, by a minute amount along the z-axis, while at the same time increasing the other dimensions so as to keep the volume ∇_t equal to 8.

Let us now consider the prism κ_h bounded by the six planes:

$$\begin{cases} x + (t - \theta)y - \alpha z + \sqrt{t}(1 + h) = 0, \\ x + (t - \theta)y - \alpha z - \sqrt{t}(1 + h) = 0, \end{cases}$$

$$\begin{cases} x + (-t - \theta)y - \alpha z + \sqrt{t}(1 + h) = 0, \\ x + (-t - \theta)y - \alpha z - \sqrt{t}(1 + h) = 0, \end{cases}$$

$$\begin{cases} z + \dfrac{2}{(1 + h)^2} = 0, \\ z - \dfrac{2}{(1 + h)^2} = 0. \end{cases}$$

Again, from formula (10.4), we find that the volume ∇_h of κ_h is

$$\nabla_h = \frac{(1 + h)^2 \left(2\sqrt{t}\right) \left(2\sqrt{t}\right) \dfrac{4}{(1 + h)^2}}{\begin{vmatrix} 1 & t - \theta & -\alpha \\ 1 & -t - \theta & -\alpha \\ 0 & 0 & 1 \end{vmatrix}} = 8.$$

Hence, there exists at least one pair of lattice points $P_2 : (p_2, q_2, r_2)$ and $P_{-2} : (-p_2, -q_2, -r_2)$ inside or on the surface of κ_h, and for such points $|r_2| = 0$ or 1.

For any value of h, where $0 < h < 1$, the total number of lattice points inside or on κ_h is finite. Now, suppose that we let $h \to 0$ so that $\kappa_h \to \kappa_t$. Then, in the limit, we have two alternative possibilities:

1. There exists at least one pair of lattice points on or inside κ_t with $|r_2| = 0$ or 1.
2. We come to a value of h, say $h = g > 0$, such that κ_g contains a pair of lattice points with $|r_2| = 0$ or 1, but also such that for all subsequent values of h (where $g > h > 0$), there are no lattice points inside κ_h with $|r_2| = 0$ or 1.

Clearly this second situation is impossible, for each prism κ_h has volume 8 and must contain nonzero lattice points; for these lattice points the only values that $|r_2|$ could have must be $|r_2| = 0$ or 1.

Going back to the prism κ_t, we can assume that for an arbitrary value of t we have a pair of lattice points P_1 and P_{-1} with $r_1 = 0$ or 1.

Assume first that for given t we have a pair of lattice points $P_1 : (p_1, q_1, 0)$ and $P_{-1} : (-p_1, -q_1, 0)$ with $r_1 = 0$. The point P_1 will lie

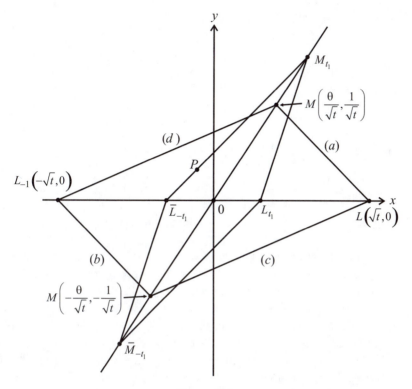

Figure 10.1

$z = 0$ plane.

inside or on the parallelogram Q_t whose sides, from (10.3), are given by the two pairs of parallel lines $(a), (b)$ and $(c), (d)$, where

$$(a): \quad x + (t - \theta)y - \sqrt{t} = 0,$$
$$(b): \quad x + (t - \theta)y + \sqrt{t} = 0,$$
$$(c): \quad x + (-t - \theta)y - \sqrt{t} = 0, \tag{10.6}$$
$$(d): \quad x + (-t - \theta)y + \sqrt{t} = 0.$$

The parallelogram so formed will resemble Figure 10.1.

The lines (a) and (c) intersect at $L : (\sqrt{t}, 0)$, while the lines (a) and (d) intersect at $M : (\theta/\sqrt{t}, 1/\sqrt{t})$. The parallelogram Q_t has *fixed* diagonals on the lines $y = 0$ and $x - \theta y = 0$. Its area A is twice the area of the triangle LMM_{-1}; that is,

$$A = 2 \cdot \frac{1}{2} \begin{vmatrix} \sqrt{t} & 0 & 1 \\ \dfrac{\theta}{\sqrt{t}} & \dfrac{1}{\sqrt{t}} & 1 \\ \dfrac{\theta}{\sqrt{t}} & -\dfrac{1}{\sqrt{t}} & 1 \end{vmatrix} = \sqrt{t} \begin{vmatrix} \dfrac{1}{\sqrt{t}} & 1 \\ -\dfrac{1}{\sqrt{t}} & 1 \end{vmatrix} + \begin{vmatrix} \dfrac{\theta}{\sqrt{t}} & \dfrac{1}{\sqrt{t}} \\ \dfrac{\theta}{\sqrt{t}} & -\dfrac{1}{\sqrt{t}} \end{vmatrix}$$

$$= \left(\sqrt{t} \right) \left(\frac{2}{\sqrt{t}} \right) + 0$$

$$= 2.$$

The point P_1 cannot lie on the diagonal $x - \theta y = 0$. For if it did, then $p_1 - \theta q_1 = 0$, or $\theta = p_1/q_1$, a rational number, which is impossible because θ is irrational. Furthermore, because $\sqrt{t} < \frac{1}{2}$, the origin is the only lattice point lying on the diagonal through $y = 0$.

Now, suppose we gradually decrease the value of t. The points M and M_{-1} will move out along the diagonal $x - \theta y = 0$, while L and L_{-1} will move in toward the origin $(0, 0)$. Eventually, we will arrive at a value $t = t_1$, for which P_1 will lie on one of the sides of $Q_{t_1} = L_{t_1} M_{t_1} L_{-t_1} M_{-t_1}$. A further decrease in t will cause P_1 to lie outside Q_t.

We cannot have two distinct pairs of lattice points in or on Q_{t_1}; otherwise its area would be greater than 2, except in the case when the vertices of Q_{t_1} are the lattice points in question. But we have just shown that we cannot have nonzero lattice points on $x - \theta y = 0$ or on $y = 0$.

We now come to the crucial part of this geometrical proof. For $t = t_1$, assume that we have a lattice point P_1 on a side of Q_{t_1}. During a further decrease in t, say from $t = t_1$ to $t = t_1 - \beta$, where $\beta > 0$, there will be no lattice points other than $(0, 0)$ in $Q_{t_1 - \beta}$. Hence, during this decrease in t, there must be a lattice point in the prism $\kappa_{t_1 - \beta}$ for which $|r_1| = 1$. Since there are only a finite number of such lattice points, it follows that at least one such point will lie on or inside $\kappa_{t_1 - \gamma}$ during a finite decrease in t, say from $t_1 - \beta$ to $t_1 - \gamma$, where $\gamma > \beta > 0$.

Thus, for some value of t such that $t_1 - \beta < t < t_1 - \gamma$, there must exist at least one lattice point $P : (p, q, 1)$ such that

$$|p - \theta q - \alpha| + t|q| < \sqrt{t}. \tag{10.7}$$

However,

$$\sqrt{|a| \cdot |b|} \leq \frac{1}{2} \{ |a| + |b| \}.$$

Hence,

$$\sqrt{t|p - \theta q - \alpha| \cdot |q|} \le \frac{|p - \theta q - \alpha| + t|q|}{2} < \frac{\sqrt{t}}{2},$$

and squaring both sides gives us

$$|p - \theta q - \alpha| \cdot |q| < \frac{1}{4}. \tag{10.8}$$

At the same time, from (10.7) and our bound on $\sqrt{t}$, we have

$$0 < |p - \theta q - \alpha| < \sqrt{t} < \epsilon. \tag{10.9}$$

Since ϵ can be made arbitrarily small, and since we assumed that $p - \theta q - \alpha \ne 0$ for any integers p, q, then (10.9) and hence (10.8) are true for an infinite number of pairs of integers (p_ϵ, q_ϵ).

Furthermore, if $q = 0$ in (10.7), then $|p - \alpha| < \sqrt{t}$. Since α is not an integer, p cannot be closer to α than one of the two integers $[\alpha]$ or $[\alpha] + 1$. Hence

$$|p - \alpha| \ge \alpha - [\alpha]$$

or

$$|p - \alpha| \ge [\alpha] + 1 - \alpha.$$

But we assumed that $\sqrt{t} < \min\{\epsilon, \alpha - [\alpha], [\alpha] + 1 - \alpha\}$. This contradiction proves that $q \ne 0$.

Finally, consider the possibility that there are no lattice points other than the origin $(0, 0)$ in the quadrilateral Q_t given by (10.6). Then there must be a lattice point $P_1(x_1, y_1, 1)$ in κ_t having $z_1 = 1$, when t lies in a certain interval as we have seen above. Again, (10.7) follows. Thus Theorem 10.2 has been proved.

Remarks. For other more difficult questions regarding this theorem, see Hardy and Wright [2].

10.3 An Application of Minkowski's Theorem

Just as Minkowski's theorems from earlier chapters give good approximations of irrational numbers by rationals, so does this last result of Tchebychev and Minkowski, but with some added strength. We let m be any positive integer and set a and b to be integers satisfying the condition that g.c.d.$(m, a, b) = 1$. Given an irrational number θ, we would like to identify a positive constant c for which infinitely many rational numbers p/q exist with the property that

$$\left|\theta - \frac{p}{q}\right| < cq^{-2}, \qquad p \equiv a \,(\mathrm{mod}\,m), \ q \equiv b \,(\mathrm{mod}\,m). \tag{10.9}$$

We denote the greatest lower bound of such numbers c by $c(\theta, m, a, b)$.

This problem was solved first by Scott [6] for $m = 2$; he found that $c(\theta, 2, a, b) \leq 1$ for arbitrary θ, a, and b. Later it was solved by Koksma [5] for general m, where it turned out that $c(\theta, m, a, b) \leq \frac{1}{4}m^2$. The proof follows directly from Theorem 10.2:

Theorem 10.3. *Let $m \geq 2$ be any integer and let a and b denote integers with* g.c.d.$(m, a, b) = 1$. *Given any irrational number θ, there exist infinitely many rational numbers p/q with*

$$\left|\theta - \frac{p}{q}\right| < cq^{-2}, \qquad \text{where } p \equiv a(\mathrm{mod}\,m), \ q \equiv b(\mathrm{mod}\,m),$$

where $c > \frac{1}{4}m^2$.

Proof. Write $p = p'm + a, q = q'm + b$. Multiplying both sides of inequality (10.9) by q^2, we get

$$|\theta q^2 - pq| < c.$$

Substituting for p and q, this becomes

$$\left|\theta(q'm + b)^2 - (p'm + a)(q'm + b)\right|$$
$$= |(q'm + b)(q'm\theta + b\theta - p'm - a)| < c.$$

Dividing the expression by m^2 and simplifying, we obtain the following:

$$\left|\left(q' + \frac{b}{m}\right)\left[q'\theta + \left(\frac{b}{m}\right)\theta - p' - \frac{a}{m}\right]\right| < cm^{-2},$$
$$\left|(q' + t)(q'\theta - p' + t\theta - s)\right| < cm^{-2},$$
$$\left|(q' + t)(q'\theta - p' - \alpha)\right| < cm^{-2}, \tag{10.10}$$

where $\alpha = s - t\theta$.

Now replacing p' and q' by their expressions in p and q, we see that $q'\theta - p' - \alpha = (q\theta - p)/m$. Because θ is irrational, this latter can never be equal to zero, so we can apply Theorem 10.2 to the last inequality above. In particular,

$$\left|(q' + t)(q'\theta - p' - \alpha)\right| = \left|q'(q'\theta - p' - \alpha) + t(q'\theta - p' - \alpha)\right|$$
$$\leq \left|q'(q'\theta - p' - \alpha)\right| + \left|t(q'\theta - p' - \alpha)\right|.$$

By the theorem,

$$\left|q'(q'\theta - p' - \alpha)\right| + \left|t(q'\theta - p' - \alpha)\right| < \frac{1}{4} + t\epsilon = \frac{1}{4} + \epsilon'$$

for any given $\epsilon' > 0$ and infinitely many rational numbers p'/q'. Thus, (10.10) will hold for $cm^{-2} > \frac{1}{4}$, proving Theorem 10.3 and showing that $c(\theta, m, a, b) \leq \frac{1}{4}m^2$.

Remarks. For this proof and many additional related results, see [1].

10.4 Proving the General Theorem

The method described above was also used by Blichfeldt to prove the more general theorem, first proved by Minkowski in 1901, namely:

Theorem 10.5. *Suppose that* $\alpha, \beta, \gamma, \delta, \xi_0, \eta_0$ *are real numbers and that* $\alpha\delta - \beta\gamma = 1$. *Then integers* p, q *always exist such that*

$$\left| (\alpha p + \beta q - \xi_0)(\gamma p + \delta q - \nu_0) \right| \leq \frac{1}{4}.$$

Proofs of this theorem have been given by Remak (1913), Mordell (1928), Landau (1931), Blichfeldt (1932), Seale (1935), and Niven (1961), to mention only a few names. See Koksma [4] for a more complete list.

References

1. P. M. Gruber and C. G. Lekkerkerker, Sections 47.3–47.6 in *Geometry of Numbers*, 2nd ed (Amsterdam and New York: North-Holland, 1987), 556–66.
2. G. H. Hardy and E. M. Wright, *An Introduction to the Theory of Numbers*, 5th ed. (Oxford: Oxford University Press, 1983).
3. Hermite, Charles, "Sur une extension donnée à la théorie des fractions continues par M. Tchebychev," *J. reine angew. Math.* 88 (1879):10–15.
4. J. F. Koksma, *Diophantische Approximationen* (New York: Chelsea, 1936).
5. ———, "Sur l'approximation des nombres irrationals sous une condition supplementaire," *Simon Steven* 28 (1951):199–202.

6. W. T. Scott, "Approximation to Real Irrationals by Certain Classes of Rational Fractions," *Bulletin of the AMS* 46 (1940):124–9.

7. P. L. Tchebychef, *Oeuvres de Tchebychef*, translated into French by A. Markoff and N. Sonin (reprinted, New York: Chelsea).

Appendix I
Gaussian Integers

Peter D. Lax

I.1 Complex Numbers

In the previous chapters we have used properties of lattice points that are related to the *additive structure* of the plane, such as what happens when sets in the plane are translated. But if we regard the plane as the plane of complex numbers, the lattice points acquire a *multiplicative structure* as well. In this appendix we shall explore this multiplicative structure.

Recall that to each point (x, y) in the plane, one can associate the complex number $z = x + iy$; x and y are called the real and imaginary parts of z, respectively. The addition of complex numbers corresponds to translations of the plane. Multiplication of complex numbers,

$$(x + iy)(u + iv) = xu - yv + i(xv + yu),$$

is related to rotation and stretching of the plane, as we shall see.

The *absolute value* of a complex number $z = x + iy$ is defined as

$$|z| = \sqrt{x^2 + y^2}.$$

According to the Pythagorean Theorem, the absolute value of z is its distance from the origin.

The *conjugate* of $z = x + iy$ is defined as $\overline{z} = x - iy$; it is the reflection of z in the real axis. There is a simple relation between $z, \overline{z}$, and $|z|$:

$$|z|^2 = z\overline{z}.$$

It follows that the absolute value of the product of two complex numbers is the product of their absolute values.

Complex numbers that correspond to lattice points,

$$g = a + ib, \qquad \text{for } a \text{ and } b \text{ real integers,}$$

are called *Gaussian integers*. Clearly the sums and products of Gaussian integers are Gaussian integers. In the language of algebra, the Gaussian integers form a *ring*. In this ring we shall investigate the same questions that we pursued in the ring of real integers: divisibility, the nature of primes, and unique factorization.

The four numbers

$$1, -1, i, -i$$

are called the *units* of the ring of Gaussian integers and are denoted by the letter u; they are the only ones that have multiplicative *inverses*. The absolute value of each unit is 1.

Problem Set for Section I.1

1. Show that no Gaussian integer that is not a unit has a multiplicative inverse in the ring of Gaussian integers.
2. Show that if $a + bi \neq \pm 1, \pm i$, then $|a + bi| > 1$.
3. Show that $(a + bi)(a - bi)$ is a real integer.

I.2 Factorization of Gaussian Integers

When a Gaussian integer g is the product of two Gaussian integers f and h, that is, when

$$g = fh, \tag{1}$$

we say that f and h *divide* g, and equation (1) is called a factorization of g, where f and h are its factors.

We can find all factors of a given Gaussian integer g by taking the absolute value in equation (1),

$$|g| = |f| \, |h|.$$

If $g = fh$, where neither f nor h is a unit, then this is called a *nontrivial factorization* of g. Since the absolute value of a Gaussian integer that is not zero or a unit is greater than one, all nontrivial factors of g have absolute value less than $|g|$. There are only finitely many such Gaussian integers f; we can check whether any of them divide g by forming the quotient

$$\frac{g}{f} = \frac{g\overline{f}}{|f|^2}$$

and seeing whether it is a Gaussian integer.

A Gaussian integer q is called *prime* if it can be factored only in a trivial way as $q = ur$, where one of the factors u is a unit. Two primes that differ by a unit factor are called *equivalent* (or *associates*).

Example 1. $5 = (2+i)(2-i)$, therefore 5 is not a prime among Gaussian integers.

Example 2. We claim that $1 + i$ is a prime. To see this, note that

$$|1+i| = \sqrt{1^2 + 1^2} = \sqrt{2}.$$

The smallest absolute value that a nonzero Gaussian integer can have is 1, and, by Problem 3 in Section I.1, the next smallest is $\sqrt{2}$. So in any factorization $1 + i = vw$, one of the factors must have absolute value 1. Therefore, that factor is a unit.

Problem Set for Section I.2

1. Show that if q is a prime, then so is $\overline{q}$.
2. Show that the primes q and $\overline{q}$, with $q \neq \overline{q}$, are inequivalent (i.e., non-associates) except for $q = (1 + i)u$.

I.3. The Fundamental Theorem of Arithmetic

Two Gaussian integers g and h are called *relatively prime* if their only common factors are units.

You will recall from Chapter 1 the Fundamental Theorem of Arithmetic for real integers: *If a and b are relatively prime integers, then 1 can be expressed as a combination of a and b of the form*

$$na + mb = 1, \quad n, \ m \text{ integers.}$$

We show now that the same result holds for Gaussian integers.

Theorem I.1. The Fundamental Theorem of Complex Arithmetic. *Let* g *and* h *be nonzero relatively prime Gaussian integers. Then* 1 *can be expressed as a combination of* g *and* h:

$$rg + th = 1, \quad r, \ t \text{ Gaussian integers.} \tag{2}$$

The proof, as in the real case, uses a form of the Euclidean algorithm. We formulate it as a lemma.

Lemma I.1.   Given two Gaussian integers g and h, we take without loss of generality

$$|g| \leq |h|.$$

Then there is a Gaussian integer f such that

$$|h - fg| < |g|. \tag{3}$$

Proof of Lemma I.1.   We shall show that, for one of the four units u,

$$|h - ug| < |h|. \tag{4}$$

This is best seen geometrically; see Figure I.1. The point h is on the circle of radius $|h|$ with center at the origin. The four points $h - g$, $h + g$, $h - ig$, $h + ig$ are located on a circle centered at h and of radius $|g|$. One of these points lies within the wedge with apex h, opening with an angle $\pi/2$ and symmetric with respect to the line through h and the origin. Call this point h_1,

$$h_1 = h - u_1 g.$$

Since $|g| \leq |h|$, we deduce from Figure I.1 that $|h_1| < |h|$; this proves that inequality (4) can be satisfied. Let P be the foot of the perpendicular from the origin, O, to a side of the wedge and let Q be the intersection of the circle of radius $|g|$ around h with the side of the wedge. Clearly $|h_1| \leq |Q|$. Right triangle QPh is isosceles, so $|h - P| = |P| = |h|/\sqrt{2}$, and

$$|h - Q| = |g|,$$

$$|P - Q| = \pm \left(|g| - \frac{|h|}{\sqrt{2}} \right).$$

By the Pythagorean Theorem,

$$|Q|^2 = \left(|g| - \frac{|h|}{\sqrt{2}} \right)^2 + \left(\frac{|h|}{\sqrt{2}} \right)^2 = |g|^2 - \sqrt{2}\,|g|\,|h| + |h|^2.$$

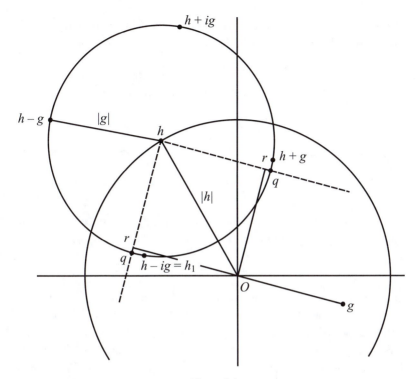

Figure I.1

Since $|g| \leq |h|$,

$$|Q|^2 \leq |h|^2 + |g|^2 - \sqrt{2}\,|g|^2 = |h|^2 - \left(\sqrt{2} - 1\right)|g|^2.$$

So

$$|h_1|^2 \leq |h|^2 - \left(\sqrt{2} - 1\right)|g|^2.$$

If $|h_1| < |g|$, we have accomplished inequality (3) with $f = u_1$; otherwise we repeat this process with h_1 instead of h. At the second step we will obtain a Gaussian integer

$$h_2 = h_1 - u_2 g = (h - u_1 g) - u_2 g = h - (u_1 + u_2)\,g = h - f_2 g.$$

Again, if $|h_2| < |g|$, we are done. If not, continuing in this fashion gives a sequence of Gaussian integers h_j, each of form $h - fg$, decreasing in absolute value at each step by a definite amount. Therefore, after a finite number of steps, we arrive at $|h_k| < |g|$. This proves the lemma.

Proof of Theorem I.1. Now, to deduce the Fundamental Theorem from the lemma, consider all Gaussian integers of form (2), that is,

$$rg + th, \quad r, t \text{ Gaussian integers.}$$

Among these, select one of smallest nonzero absolute value; denote it by s:

$$s = rg + gh.$$

Since g and h are not zero and are of the form (2), and since s is the one with smallest positive absolute value, it follows that

$$|s| \leq |g|, \quad |s| \leq |h|.$$

Applying the lemma to h and s, we conclude that there is a Gaussian integer f such that $|h - fs|$ is less than $|s|$. But $h - fs$ is also of form (2); therefore by the minimum property of s, $h - fs = 0$. In words: s divides h. Similarly we deduce that s divides g; since h and g are relatively prime, their only common divisors are the units; therefore s is a unit. This proves that one of the units can be represented in form (2). But then so can the unit 1:

$$rh + tg = 1. \tag{5}$$

This completes the proof.

Note that our proof that (5) has a solution is nonconstructive.

Problem for Section I.3

1. Use Lemma I.1 to fashion a Euclidean algorithm for the construction of a solution of equation (5).

I.4. Unique Factorization of Gaussian Integers

The Fundamental Theorem of Complex Arithmetic has the same consequences about divisibility and factoring of Gaussian integers as its counterpart has for real integers.

Theorem I.2. *If g and h are relatively prime Gaussian integers, and g divides hk, then g divides k.*

Proof. Since g and h are relatively prime, by the Fundamental Theorem there are Gaussian integers r and t such that

$$rg + th = 1.$$

Multiply this relation by k:

$$rgk + thk = k.$$

By hypothesis, g divides hk, so both terms on the left are divisible by g; therefore, so is the right side. The proof is complete.

The following theorem is an immediate consequence of Theorem I.2.

Theorem I.3. *Let q be a Gaussian prime, and suppose that q divides the product fg. Then q divides f or g.*

Proof. Suppose that q does not divide f; then q is relatively prime to f. So, by Theorem I.2, q divides g, and the proof is complete.

Problem for Section I.4

1. Show that every Gaussian integer can be written as a product of primes, and that the primes appearing in this factorization are uniquely determined except for unit factors.

I.5.   The Gaussian Primes

This brings us to the natural problem of describing all Gaussian primes. Here is the answer, which has four parts.

Theorem I.4.

a. *Every prime p in the ring of real integers of form $4n + 3$ is a Gaussian prime.*

b. *All real primes p of form $4n + 1$ can be factored, essentially uniquely, as $p = q\bar{q}$, where q is a Gaussian prime.*

c. *The number* 2 *is not a Gaussian prime; it can be factored as* $2 = i(1 - i)^2$.

d. *All the Gaussian primes are accounted for in parts a, b, or c.*

Proof. We begin by proving part a.

If a real prime p of form $4n + 3$ were not a Gaussian prime, it would have a nontrivial factorization $p = st$, where s and t are Gaussian integers and neither is a unit. Take the absolute value squared of this product:

$$p^2 = |s|^2 |t|^2.$$

Since p is a real prime, the only nontrivial factorization of p^2 is $p \cdot p$. Therefore

$$|s|^2 = |t|^2 = p.$$

We can express s as $a + ib$, for a, b real integers. The above relation implies that

$$a^2 + b^2 = p.$$

The square of a real integer is congruent either 0 or 1 mod 4; therefore the sum of the squares of two real integers is never of the form $4n + 3$. This contradiction proves part a of Theorem I.4.

Part b of Theorem I.4 deals with real primes of form $4n + 1$. For this we need Wilson's Theorem.

Theorem I.5 (Wilson). *Let p be any real prime. Then $(p - 1)!$ is $\equiv -1$ mod p.*

Proof of Wilson's Theorem. Denote by $R(p)$ the collection of congruence classes mod p. $R(p)$ consists of p elements, represented for instance by the residues $0, 1, 2, \ldots, p - 1$. Since sums and products of congruent numbers are congruent, addition and multiplication are defined in $R(p)$, making it a commutative ring. We claim that $R(p)$ is a *field*: that is, every nonzero element of $R(p)$ has a multiplicative inverse.

This is not hard to show. Take any number a not divisible by p. We claim that no two of the numbers $a, 2a, \ldots, (p - 1)a$ are $\equiv$ mod p. To see this, take $r \not\equiv s$; we claim that $ar \not\equiv as$. For, their difference $ar - as$ can be factored as $a(r - s)$; since neither a nor $r - s$ is divisible by p, their product is not divisible by p either. Since the $p - 1$ numbers $a, 2a, \ldots, (p - 1)a$

are not divisible by p, they can be put into one-to-one correspondence with the $p - 1$ residues $1, 2, \ldots, p - 1$, so that corresponding numbers are $\equiv$ mod p. In particular, there is an r such that $ar \equiv 1$ mod p. This r is the multiplicative inverse of a.

Some congruence classes are their own inverses. For instance, $1 \cdot 1$ and $(p - 1) \cdot (p - 1)$ are congruent 1 mod p. No other class is its own inverse; that is, $r^2 \not\equiv 1$ mod p unless $r \equiv 1$ or -1. This is easy to see: $r^2 - 1 = (r - 1)(r + 1)$ is not divisible by p unless either $r - 1$ or $r + 1$ is.

We are now ready to prove Wilson's Theorem. In the product

$$(p - 1)! = 1 \cdot 2 \cdot \cdots \cdot (p - 1),$$

pair each factor, except the first and the last, with its multiplicative inverse. The product of such a pair is $\equiv 1$ mod p; this proves that $(p - 2)! \equiv 1$ mod p. Multiplying this relation by $p - 1$ shows that $(p - 1)! \equiv -1$ mod p, as asserted in Wilson's Theorem.

Continuing with our proof of Theorem I.4, part b, we now factor $(p - 1)!$ as follows:

$$(p - 1)! = \left[1 \cdot 2 \cdot \cdots \cdot \frac{p - 1}{2} \right] \left[(p - 1) \cdot (p - 2) \cdot \cdots \cdot \frac{p + 1}{2} \right] = fg.$$

Each factor f and g is the product of $(p - 1)/2$ factors, and the jth factor of g is congruent to the negative of the jth factor of f. It follows that $g \equiv (-1)^{(p-1)/2} f$. When p is of form $4n + 1$, then $(p - 1)/2$ is even; so in this case $f \equiv g$ mod p. Using Wilson's Theorem, we conclude from the above that $f^2 \equiv -1$ mod p, which means that $f^2 + 1$ is divisible by p.

The next step comes from the notes of the great geometer János Bólyai [see Elemér Kiss, "Fermat's Theorem in János Bólyai's Manuscripts," *Mat. Pannonica* 6 (1995):344–8]. Bólyai factors $f^2 + 1$ in the ring of Gaussian integers as $(f + i)(f - i)$. As we have shown, this product is divisible by p. If p were a Gaussian prime, it would divide one of the factors $f + i$ or $f - i$; neither is possible, for it would mean that

$$f \pm i = p(a + ib),$$

making $pb = \pm 1$, which is absurd. So p is not a Gaussian prime.

Let q be a Gaussian prime that divides p: $p = qw$. Taking complex conjugates gives $p = \overline{q}\,\overline{w}$, which shows that also $\overline{q}$ divides p. Since q and $\overline{q}$ are distinct primes, it follows from Theorem I.3 that also their product

$q\bar{q} = |q|^2$ divides p. Since p is a real prime, $|q|^2 = p$. This completes the proof of part b.

Now we turn to part d of Theorem I.4. To prove that all nonreal Gaussian primes q are as described in part b or part c, we will show that $|q|^2$ is a real prime. Denote $|q|^2$ by c: $c = q\bar{q}$. If c were not a prime, it could be factored into $c = ab$, for a and b real integers, both greater than 1. We would then have $ab = q\bar{q}$. This shows that q divides ab. According to Theorem I.3, the prime q divides a or b; say $a = qf$. Since q is not real, f is not a unit. Setting this into the previous formula gives $qfb = q\bar{q}$. Dividing both sides by q, we get $fb = \bar{q}$, a nontrivial factorization of $\bar{q}$. This is impossible because $\bar{q}$ is a Gaussian prime.

Once we know that $q\bar{q}$ is a real prime, we are back in cases b or c. This completes our proof of Theorem I.4.

I.6.   More about Gaussian Primes

Theorem I.4 has interesting consequences, as we shall see next.

Corollary 1.   A real prime p of form $4n + 1$ can be written as a sum of squares of two real integers $p = a^2 + b^2$ in essentially one way. That is, if $p = a^2 + b^2 = a_1^2 + b_1^2$, then $a + bi$ and $a_1 + b_1 i$ differ by a unit.

As an exercise, the reader may prove Corollary 1.

Corollary 2.   The positive integer m can be written as a sum of squares of two real integers,

$$m = a^2 + b^2, \tag{6}$$

if and only if all real prime divisors of m of form $4n + 3$ divide m with even multiplicity.

Proof.   Factor m as $m = st$, where s is the product of all real prime divisors of m of form $4n + 3$, and t is the product of all the others. If each prime factor of s occurs an even number of times, s is a perfect square, and so we can write $m = r^2 t$, for r a real integer.

Each real prime factor p_j of t either is 2 or is of the form $4n + 1$. Therefore, according to Corollary 1, it can be represented as $p_j^2 = a_j^2 + b_j^2$, for a_j and b_j real integers. The product of two sums of two squares can be written as a sum of two squares:

$$(a^2 + b^2)(c^2 + d^2) = (ac - bd)^2 + (ad + bc)^2.$$

So we conclude recursively that the product $t = \prod p_j$ is a sum of the squares of two integers. But, then, so is $r^2 t = st = m$.

To show that the condition imposed on s is necessary, assume that m can be written in form (6). In the ring of Gaussian integers, (6) can be factored as

$$m = (a + ib)(a - ib).$$

Suppose that a real prime p of form $4n + 3$ divides m. According to Theorem I.4, p is a Gaussian prime as well; according to Theorem I.3, p divides $a + ib$ or $a - ib$. Let's say

$$a + ib = pw. \tag{7}$$

Taking complex conjugates, we get

$$a - ib = p\overline{w}. \tag{7'}$$

This shows that p divides both, $a + ib$ and $a - ib$, so we conclude that m is divisible by p^2. Proceeding recursively, we remove pairs of prime factors of form $4n + 3$. In a finite number of steps we will have removed them all, and since they go in pairs, each occurs with even multiplicity as asserted in our statement of Corollary 2.

We observe, furthermore, that if a and b in representation (6) are relatively prime, then m is not divisible by any prime of the form $4n + 3$. For, adding and subtracting equations (7) and (7'), we would conclude that p divides both a and b, contrary to assumption.

Examples of primes of the form $4n + 3$ are $3, 7, 11, 19, 23, \ldots$; on the other hand $5, 13, 17, 29, \ldots$ are primes of the form $4n + 1$. We conclude this chapter by showing that there are infinitely many primes of each kind. The case of primes of form $4n + 3$ can easily be handled by the following slight modification of Euclid's classic argument. Suppose to the contrary, that there were only a finite number of such primes $p_1, p_2, \ldots, p_k$. Form the product $s = 4 \prod p_j$. Clearly, $s - 1 \equiv 3 \bmod 4$. Factor $s - 1$ into primes over the real integers; at least one of these prime factors must be of form $4n + 3$, otherwise their product would be $\equiv 1 \bmod 4$. But this prime factor of $s - 1$ is relatively prime to s, so it would be different from all the p_j, where $j = 1, 2, \ldots, k$.

Suppose now that there were only a finite number of primes $p_1, \ldots, p_k$ of form $4n + 1$. Form their product $a = \prod p_j$ and define

$$m = a^2 + 1.$$

Since a and 1 are relatively prime, it follows from our observation above that all real prime factors of m are of form $4n + 1$. None divide a^2, so they must be different from all the p_j, where $j = 1, \ldots, k$.

Appendix II
The Closest Packing of Convex Bodies

As mentioned in the Preface, the lattice-point problems central to the geometry of numbers have extensive connections to modern mathematics and its applications. They arise in the theories of finite groups, of quadratic forms, of combinatorics, and of numerical methods for evaluating n-dimensional integrals. They appear in chemistry and physics, in crystallography in particular, and in the design of codes for transmitting, storing, and receiving data. In this appendix, we give a brief introduction to sphere packing, which, in turn, is critical to the development of error-detecting and error-correcting codes. In fact, finding dense packings of spheres into a given space is a problem equivalent to that of finding efficient error-correcting codes. Though our discussion here is brief, no introduction to the geometry of numbers could be complete without giving the reader at least a glimpse into this important and timely application.

II.1   Lattice-Point Packing

Let K be a convex set, or body, symmetrically placed about the origin, O. Suppose that we have an *admissible lattice for* K; that is, one that has no lattice point inside K other than the origin. If K is shrunk to half its linear dimensions, to $\frac{1}{2}K$, and if this body is then translated to have its center at each lattice point, the resulting bodies will not overlap. Conversely, if the lattice had a point other than O inside K, the resulting bodies would overlap. Thus, an admissible lattice for K means precisely a lattice that provides a non-overlapping packing for the convex body $\frac{1}{2}K$.

Now let us consider the density of such a packing in an arbitrary admissible lattice, that is, the proportion of space occuped by the translates of $\frac{1}{2}K$. Denote by $V\left(\frac{1}{2}K\right)$ the volume of each of the translates. In a large cube of volume V, the number of lattice points is asymptotic to V/Δ, where Δ is the volume of the fundamental domain of the admissible lattice under consideration. (Recall that this volume is given by $|\Delta| = |ad - bc|$ when the lattice is generated by the vectors (a, c) and (b, d).) Hence the total volume of the tranlates enclosed by V, each centered at a lattice point, is $V\left(\frac{1}{2}K\right)V/\Delta$ and the density of the packing is given by

$$\frac{V\left(\frac{1}{2}K\right)}{\Delta} = \frac{V(K)}{2^n\Delta}.$$

This density is greatest when Δ is as small as posible. If K is a set with at least one admissible lattice, we take

$$\Delta(K) = \inf \Delta(L)$$

where the infimum (or greatest lower bound) is taken over all admissible lattics for K. The number $\Delta(K)$ is called the *critical determinant of K*, and any lattice with $\Delta(L) = \Delta(K)$ is called a *critical latttice of K*. So, if we let $\delta(K)$ represent the *density of the closest lattice packing of K*, then

$$\delta(K) = \frac{V(K)}{2^n\Delta(K)} \leq 1.$$

The number $\delta(K)$ is unaltered by any linear transformation since both $V(K)$ and $\Delta(K)$ are multiplied by the same factor under such a transformation.

There exist bodies for which $\delta(K) = 1$. The unit cube in $\mathbb{R}^n$ is one example, while the regular hexagon in $\mathbb{R}^2$ is another. Copies of such bodies can be arranged without overlapping interiors so that they completely fill a given space; the centers of each translate then form a lattice in the sense that if u and v are centers, then so are $u + v$ and $u - v$.

II.2   Closest Packing of Circles in $\mathbb{R}^2$

Unlike squares or hexagons, circles cannot be arranged to completely fill an arbitrary region in the plane unless we allow them to overlap. Suppose that we do not permit such overlaps, but instead insist that any circle must touch another in, at most, a single tangent point. How efficiently might they be packed together now?

Given a unit circle centered at the origin, its circumference will, of course, pass though the point $(1, 0)$. We consider any admissible lattice for this closed unit disk, D, that contains the point $(1,0)$. Then the x-axis will contain the lattice points $(n,0)$, where n is any integer; all other lattice points will be contained on lines parallel to the x-axis and lying above or below it. The nearest such line of lattice points must be at a perpendicular distance of at least $\sqrt{3/2}$ units from the x-axis. Indeed, suppose that (p,q) is a lattice point with $p > 0$ and $q < \sqrt{3/2}$. Consider the lattice point

$$(p,q) - ([p],0) = (p - [p], q),$$

which is either the point $(0,q)$ or lies inside the vertical strip bounded by the lines $x = 0$ and $x = 1$. If $0 \le p - [p] \le \frac{1}{2}$, we let $p_1 = p - [p]$; if $\frac{1}{2} < p - [p] < 1$, we let $p_1 = p - [p] - 1$. In either case, we have $|p_1| \le \frac{1}{2}$. But then

$$p_1^2 + q^2 < \left(\frac{1}{2}\right)^2 + \left(\frac{\sqrt{3}}{2}\right)^2 = 1,$$

showing that (p_1, q) is a lattice point contained in the interior of the disk, contradicting the definition of an admissible lattice for D.

Suppose we consider a lattice generated by the two vectors $(1,0)$ and $\left(\frac{1}{2}, \frac{\sqrt{3}}{2}\right)$. This lattice is clearly admissible for D, since it provides a packing for $\frac{1}{2}K$. Furthermore, L. Fejes Toth [6] proved in 1940 that it is critical and, hence, that $\Delta(D) = \sqrt{3}/2$. Note that this critical lattice is hexagonal, with the vertices of the hexagon at $(\pm 1, 0)$, $\left(\pm\frac{1}{2}, \pm\frac{\sqrt{3}}{2}\right)$. The density of the closest lattice packing for D is therefore,

$$\delta(D) = \frac{v(D)}{4\Delta(D)} = \frac{\pi}{4\left(\frac{\sqrt{3}}{2}\right)} = \frac{\pi}{2\sqrt{3}} = 0.90668996\ldots.$$

Interest in an optimal density for two dimensions is well founded. Its solution is important to many practical design issues such as determining how best to place wires in a cross-section of cable. For circles in $\mathbb{R}^2$, much more can be proved. We refer the interested reader to B. Segre and K. Mahler's article, "On the Densest Packing of Circles" [14].

II.3   The Packing of Spheres in $\mathbb{R}^n$

A system of spheres of equal volume, V, in $\mathbb{R}^n$ is said to form a packing if no two spheres of the system have an interior point in common. In the previous section, we constructed a critical lattice for a system of unit circles, where $n = 2$, and found its density. For $n = 3$, the problem has an interesting history. It is related by Philip Griffiths in the *American Mathematical Monthly* [**8**] and by Thomas Hales in the *Notices of the AMS* [**11**]. We pass on a brief version here.

In the late 1500s Sir Walter Raleigh posed the question of how to most efficiently stack cannonballs on the deck of a ship. The English mathematician Thomas Harriot, to whom he had taken his inquiry, wrote in turn to the astronomer Johannes Kepler about the problem. Kepler conjectured that the densest packing was to be found in the usual way that sailors had then of stacking cannonballs and that grocers have now of stacking oranges. This became known as the Kepler conjecture, and it remained an important outstanding problem for many years, in spite of the vast amount of mathematical research it inspired.

At any greengrocer's, we can see a rectangular base of fruit, with successive layers stacked into natural "deep holes" left by the lower layer, ultimately forming a pyramid-like structure. The centers of each fruit (or cannonball) resulting from such a packing form a lattice, called the *face-centered cubic lattice* (or fcc lattice, for short). They can be described very simply as the set of lattice points (x, y, z) in $\mathbb{Z}^3$ with $x + y + z$ even. The density of this packing is

$$\delta(K) = \frac{\pi}{\sqrt{18}} = 0.7405\ldots.$$

Thus, the usual version of the sphere-packing problem asked if this density were maximal among all packings. Gauss proved in 1831 [**7**] that the fcc lattice is critical; that is, the fcc lattice gives the densest among lattice packing s.

The problem is complicated, however, by the fact that not all packings are given by lattices, and the general question proved much harder to resolve. L. Fejes Toth made major progress when in 1953 he reduced the problem to an extremely large calculation that might be solved by computer. The best bound known for many years was given by C. A. Rogers [**13**] in 1958, who showed that no sphere packing could have density greater than 0.7796 . . . . Not long ago, the truth of the Kepler conjecture was proved by Hales, who used recent mathematical advances as well as vast computing power to prove his result [**9, 10**]. His abovementioned survey article [**11**] gives a good overview of ths history of the problem and an accessible description of his work.

Let us now consider, for $n = 4$, the *checkerboard lattice*. Like the fcc lattice, the checkerboard lattice is defined as the set of points (x, y, z, w) in $\mathbb{Z}^4$ with $x + y + z + w$ an even integer. A set of generating vectors is given by $(2, 0, 0, 0)$, $(1, 1, 0, 0)$, $(1, 0, 1, 0)$, and $(1, 0, 0, 1)$, so $\Delta = 2$, while the spheres are taken to have radius $\sqrt{2}/2$. For this lattice, therefore, we have the density

$$\delta(K) = \frac{\pi^2}{16} = 0.6169\ldots.$$

Korkine and Zolotareff [12] proved in 1872 that the checkboard lattice is critical and, hence, that $\pi^2/16$ is the maximal density among lattice packings. For this and many additional results in higher dimensions, the reader should consult Conway and Sloane's 1988 book, *Sphere Packings, Lattices and Groups* [5], which gives an excellent simple introduction to sphere packing and a complete summary of what was known up to that time. In addition, we refer the interested reader to Thompson's book, *From Error-Correcting Codes through Sphere Packings to Simple Groups* [15], which gives a good summary of known results, as well as a clear presentation of the connections among those three topics and their extremely interesting history.

Blichfeldt, in fact was the first to obtain an upper bound less than 1 for the density of the closest sphere packing in arbitrary dimensions. This was published in his famous paper of 1914 [1], but his proof related only to lattice packings. He proved the following theorem.

Theorem 10.4 (Blichfeldt). *The density of the closest packing of equal spheres in $\mathbb{R}^n$ does not exceed*

$$\delta(K) \leq \frac{n + 2}{\left(\sqrt{2}\right)^{n+2}}.$$

Blichfeldt later proved that certain lattices were critical in six, seven, and eight dimensions [2–4]. In [3], he improved the result of Theorem 10.4 to apply to arbitrary packings.

References

1. H. F. Blichfeldt, "A New Principle in the Geometry of Numbers with Some Applications," *Transactions of the AMS* 15:3 (July 1914):227–35.
2. _____, "On the Minimum Value of Positive Real Quadratic Forms in 6 Variables," *Bulletin of the AMS* 31 (1925):386.
3. _____, "The Minimum Value of Quadratic Forms, and the Closest Packing of Spheres," *Mathematische Annalen* 101 (1929):605–8.

4. _____, "The Minimum Values of Positive Quadratic Forms in Six, Seven, and Eight Variables," *Mathematische Zeitschrift* 39 (1934):1–15.

5. J. G. Conway and N. J. A. Sloane, *Sphere Packings, Lattices and Groups* (New York and Berlin: Springer-Verlag, 1988), Preface and Chapter 1, 1–30.

6. L. Fejes Toth, "Über Einen geometrischen Satz," *Mathematische Zeitschrift* 46 (1940):79–83.

7. C. F. Gauss, "Besprechung des Buchs von L. A. Seeber: Untersuchungen über die Eigeschaften der positiven ternaren quadratischen Formen usw.," *Göttingsche Gelehrte Anzeigen* (July 9, 1831); reprinted in *Werke*, Vol. 2 (Göttingen: Gesell-schaft der Wissenschaften, 1876), 188–96.

8. Phillip A. Griffiths, "Mathematics at the Turn of the Millennium," *American Mathematical Monthly* 107 (2000):1–14.

9. Thomas Hales, "Sphere Packings. I," *Discrete Comp. Geom.* 17 (1997):1–15.

10. _____, "Sphere Packings. II," *Discrete Comp. Geom.* 18 (1997):135–49.

11. _____, "Cannonballs and Honeycombs," *Notices of the AMS* 47 (2000):440–9.

12. A. Korkine and E. I. Zolotareff, "Sur les formes quadratiques positives quaternaires," *Mathematische Annalen* 5 (1872):581–3.

13. C. A. Rogers, "The Packing of Equal Spheres," *Proceedings of the London Mathematical Society* 8 (1958):609–20.

14. B. Segre and K. Mahler, "On the Densest Packing of Circles," *American Mathematical Monthly* 51 (1944):261–70.

15. Thomas M. Thompson, *From Error-Correcting Codes through Sphere Packings to Simple Groups*, Carus Monograph Series, No. 21 (Washington, DC: MAA, 1983).

Appendix III
Brief Biographies

Hermann Minkowski (1864–1909)

"Rien n'est beau que le vrai, le vrai seul est amiable."
—Minkowski's motto

Born in Alexoten, Russia, in 1864, Hermann Minkowski was brought up in Königsberg, Germany, where he also spent most of his university years. Rising rapidly through the ranks of academia, he earned the title Dokter in 1885, became a Docent at the University of Bonn in 1887, and in 1892 was promoted to Extraordinarius. During vacations, Minkowski usually returned to Königsberg to work with David Hilbert (1862–1943) and Adolf Hurwitz (1858–1919). For a time he actually resettled in Königsberg, where the university made him Extraordinarius in 1894 and Ordinarius (something like a full professor) in 1895. However, he was lured to Zurich, Switzerland, becoming an Ordinarius at the Polytechnicum in 1896 and marrying the following year. Eventually, to be with Hilbert again, Minkowski moved to Göttingen in 1902. It was there he died unexpectedly in 1909 of acute appendicitis. Only forty-five years old, Minkowski was struck down in the prime of his scientific career.

Minkowski had been a precocious genius. Upon completing his gymnasium education in Königsberg, he entered the university in 1880—not quite sixteen years old! During his first five semesters, Minkowski studied under H. Weber and W. Voigt. Then, going to Berlin for three semesters, he took courses from E. Kummer, Leopold Kronecker, Karl Weierstrass, H. L. F. von Helmholtz, and R. Kirchhoff. Some of these names reveal that Minkowski was a serious student of physics. This readily explains his later fascination with relativity theory.

In 1881, the French Academy proposed for its Grand Prix a difficult problem related to quadratic forms, with participants submitting entries anonymously under code names. The solution involved demonstrating and completing certain theorems of F. Eisenstein (1823–1852), who had been a pupil of Gauss's and a favorite member of that select group. Eisenstein had died at barely twenty-nine years old. Incredibly, Minkowski himself was not yet eighteen when he completed his solution of the Academy's problem, and he was awarded the prize a year later, in 1883.

He was not alone, for the Grand Prix was awarded jointly that year, going posthumously to the English mathematician H. J. S. Smith (1826–1883), one month after his death. Smith was famous for having spent ten years studying nearly everything of importance in the theory of numbers, summarizing it all in his *Reports*, which appeared in the *British Association* volumes from 1859 to 1865. These reports still stand as models of clear and precise exposition. Smith also contributed two memoirs to the *Proceedings of the Royal Society* in 1864 (the year of Minkowski's birth!) and 1868, wherein he had already solved the problem proposed in 1881 by the French Academy. Smith sent the Academy his dissertation in 1882, and it was decided that a joint award was merited.

Unfortunately, the Academy's decision had some unpleasant results for Minkowski. So great was the prestige of the Grand Prix, and so strong was the anti-German sentiment after the Franco–Prussian war, that some scientists decried the joint award, launching attacks on Minkowski in the press and making unfounded suggestions of plagiarism. To their credit, such *savants grands* of the French Academy as Camille Jordan, J. Bertrand, and Charles Hermite stood up firmly in Minkowski's defense. Writing directly to him at this unhappy time, Jordan was warmly encouraging: "Work hard, young man; I beg you to become an eminent geometer!"

As mentioned, Minkowski was talented in physics as well as mathematics. When a thirty-page paper was published in 1905 in the *Annallen der Physik* (vol. 17) by the largely unknown Albert Einstein, Minkowski studied it closely. Eventually, starting from the principles stated by Einstein, Minkowski made his own important contributions to the special theory of relativity. In 1908, Minkowski gave a remarkable lecture entitled *Raum und Zeit*, offering a new view of space and time. One of Minkowski's fundamental points was that mass and energy are proportional. Later, Minkowski

generalized this theory, coming to the conclusion that a ray of light is at-
tracted by matter.

Soon after, he lay dead—a great and active mind now stilled. Who
knows what his enormous talents might have accomplished in the new
realms to which physics was turning? In mathematics, at least, his legacy
is clear: Minkowski's beloved "geometry of numbers" survives him.

Hans Frederik Blichfeldt (1873–1945)

Hans Frederik Blichfeldt was born in 1873 in a small village in Grønbeck
Sogn, Denmark. Earlier Blichfeldts included farmers, ministers, and bish-
ops; on his mother's side of the family there were many scholars and
teachers. Before Blichfeldt emigrated to the United States in 1888 with his
father and an older half-brother, he had passed with high honors the state
examinations conducted by the University of Copenhagen. By the time he
took this examination, he had already discovered by himself the solution
of the general polynomial equations of the third and fourth degrees—a
remarkable performance for someone not yet fifteen years old.

As a young man, Blichfeldt had a sturdy physique. This served him
well during his first four years in the United States, where he "worked
with his hands doing everything," mostly in the lumber industries of the
Pacific Northwest. From 1892 to 1894 he worked as a draftsman for the
city and county of Whatcom, Washington, where his unusual mathematical
talent came to the attention of his engineer employers, who persuaded
him to apply for admission to the recently opened Stanford University in
California.

Blichfeldt was admitted to Stanford in September 1894, and by June
1896 had his A.B., followed by the A.M. degree in 1897. The "free elec-
tive" system then in effect at Stanford allowed Blichfeldt to concentrate
on mathematics and arrive quickly at his goals.

In the late nineteenth century, it was the custom for aspiring young
mathematicians to go to Germany for advanced instruction. Blichfeldt was
determined to attend the University of Leipzig and study under Sophus Lie
(1842–1899). Despite his thrift, he had to borrow money to go abroad.

Blichfeldt spent the year 1897–1898 working under Lie, mastering
the "Lie Theory" of continuous groups, and in 1898 he received his Ph.D.
summa cum laude. His doctoral dissertation, "On a Certain Class of Group
Transformations in Space of Three Dimensions," was published in 1900

in the *American Journal of Mathematics* (vol. 22, pp. 113–20). Returning to Stanford, Blichfeldt taught there for the next forty years. He became a professor in 1913 and was Head of the Mathematics Department from 1927–1938.

Blichfeldt published relatively few papers, perhaps some twenty-five research articles. He was the author of two books, *Finite Groups of Linear Homogeneous Transformations* and *Finite Collineation Groups*.

The foregoing sketch does not adequately convey Blichfeldt's potentialities. He attacked many of the difficult problems of his day, disposing of some, but leaving many more endeavors in incomplete form. Too often, once he had reasoned a problem through to his satisfaction, the drudgery of reducing his notes for publication was too much for him to undertake. One expert commented that Blichfeldt left behind "a barrel of good stuff." Many times he simply published a brief abstract of his results in the *Bulletin of the American Mathematical Society*, and let it go at that. Two works that are of special concern to the readers of this book are his 1914 paper ("A New Principle in the Geometry of Numbers with Some Applications") and his frequently cited 1934 study ("The Minimum Values of Positive Quadratic Forms in Six, Seven, and Eight Variables"), mentioned in Chapters 8 and 9, respectively.

His distinctions were many. They include vice-presidency of the American Mathematical Society (1912); election to the National Academy of Science in 1920; membership in the National Research Council from 1924 through 1927; and investiture as Knight of the Order of Dannebrog (Denmark), in 1939.

References

1. Harold M. Bacon, *Dictionary of American Biography, Third Supplement, 1941–45* (New York: Scribner, 1973); see Blichfeldt.
2. E. T. Bell, "Hans Frederik Blichfeldt," *National Academy Biographical Memoirs*, Vol. 26, 181–7.
3. _____, *Development of Mathematics*, reprint of 2nd ed. (New York: Dover, 1992); re: Minkowski.
4. Florian Cajori, *A History of Mathematics* (New York: Macmillan, 1931).
5. J. Fang, *Hilbert* (New York: Paideia Press, 1970); re: Minkowski.
6. David Hilbert, "Hermann Minkowski," *Math. Annalen* 68 (1910):455–71.

Solutions and Hints

Chapter 1

1 a. $y = \frac{2}{3}x + \frac{1}{5}$ (because 5 does not divide 3). Other examples are similar.

  b. $y = \frac{2}{15}x + \frac{1}{3}$ (because $3 \mid 15$). $(p_0, q_0) = (5, 1)$; $\{(p_k, q_k) = (5 + 15k, 1 + 2k), \mid k$ is any integer$\}$.

2. *Hint:* Use the distance formula to find the length d_k of the line segment from (p_{k-1}, q_{k-1}) to (p_k, q_k). If d_k is not a function of k, then d_k is constant from any point to any adjacent lattice point.

3. *Hint:* Let $x = p_1 + k(p_2 - p_1)$ and $y = q_1 + k(q_2 - q_1)$. Use the fact that $q_1 = mp_1 + b$ and $q_2 = mp_2 + b$.

4. Clearly $(p, q) = (n, m)$ lies on the line, and n and m are relatively prime. $|(n, m)| = \sqrt{n^2 + m^2}$. If (p_1, q_1) lies on the line, then $mp_1 = nq_1$. Since g.c.d.$(n, m) = 1$, $n \mid p_1$, say $nr_1 = p_1$. Then $q_1 = mr_1$, so $|(p_1, q_1)| = \sqrt{n^2 r_1^2 + m^2 r_1^2} = r_1\sqrt{n^2 + m^2} \geq |(n, m)|$.

5. No. If (p, q) lies on the line, then $\sqrt{2} = q/p$.

6 a. Let $p = q = 1$. $\left|\sqrt{2}p - q\right| < |1.4142136(1) - 1| = 0.4142136 < \epsilon = \frac{1}{2} = 0.5$.

  b. Let $p = 10, q = 14$. $\left|\sqrt{2}(10) - 14\right| < |14.142136 - 14| = 0.142136 < \epsilon = \frac{1}{5} = 0.2$.

  c. Let $p = 5, q = 7$. $\left|\sqrt{2}(5) - 7\right| = \frac{1}{2}\left|\sqrt{2}(10) - 14\right| < \left(\frac{1}{2}\right)\frac{1}{5} = \frac{1}{10} = 0.1$.

Section 2.1

1 a. $[1.3 + 2.8] = [4.1] = 4$, while $[1.3] + [2.8] = 1 + 2 = 3$.

  b. $\left[\frac{5.4}{2.7}\right] = [2] = 2$, while $\frac{[5.4]}{[2.7]} = \frac{5}{2} = 2.5$.

  c. $[(3.7)(2.6)] = [9.62] = 9$, while $[3.7][2.6] = (3)(2) = 6$.

2. Let $x = [x] + \zeta$, where $0 \leq \zeta < 1$. $[x + n] = [[x] + \zeta + n] = \left[([x] + n) + \zeta\right] = [x] + n$.

3. *Hint:* $[n] = n$ and $[-n] = -n$ if n is an integer. $x = [x] + \zeta$ and $-x = -[x] - \zeta$ if x is not an integer. Thus $[-x] = -[x] - 1$.

4. $[x/n] = \left[\left([x] + \zeta\right)/n\right] = \left[([x]/n) + (\zeta/n)\right]$, where $\zeta/n < 1/n$. $[x]/n = q + (r/n)$, where $0 \le r \le n - 1$. Thus,

$$q < \left(\frac{[x]}{n}\right) + \left(\frac{\zeta}{n}\right) < q + \frac{n-1}{n} + \frac{1}{n} = q + 1, \quad \text{and}$$

$$\left[\frac{x}{n}\right] = \left[\frac{[x]}{n} + \frac{\zeta}{n}\right] = q.$$

5. $[2x] + [2y] = \left[2([x] + \zeta_1)\right] = \left[2([y] + \zeta_2)\right] = 2[x] + [2\zeta_1] + 2[y] + [2\zeta_2]$. $[x] + [y] + [x + y] = [x] + [y] + \left[[x] + \zeta_1 + [y] + \zeta_2\right] = 2[x] + 2[y] + [\zeta_1 + \zeta_2]$. If $0 \le \zeta_1, \zeta_2 < \frac{1}{2}$, then $0 \le 2\zeta_1, 2\zeta_2 < 1$, so $[2\zeta_1] = 0 = [\zeta_1 + \zeta_2]$. If $0 \le \zeta_1 < \frac{1}{2}, \frac{1}{2} \le \zeta_2 < 1$, then $[2\zeta_1] = 0$, $[2\zeta_2] = 1$, $[\zeta_1 + \zeta_2] = 0$ or 1. If $\frac{1}{2} \le \zeta_1, \zeta_2 < 1$, then $[2\zeta_1] = [2\zeta_2] = 1 = [\zeta_1 + \zeta_2]$.

6. If $a < b$, then $0 \times b < a < 1 \times b$, while $[a/b] = 0$. If $a = b$, then $1 \times b = a$, while $[a/b] = 1$. If $a > b$, then divide the positive real line into adjoining intervals of length b. Since a must lie in some interval, $kb < a < (k + 1)b$ for some integer k. Then $a = kb + r$, so $[a/b] = [k + (r/b)] = k$.

7. *Hint:* Make a graph of the line showing x, $[x]$, $-[x]$, and $[-x]$. Note then that $-[-x] = -\left[-[x] - \zeta\right] = -(-[x]) - [-\zeta] = [x] - (-1) = [x] + 1$.

8. If $x = n$, an integer, then $\left[x + \frac{1}{2}\right] = \left[n + \frac{1}{2}\right] = n = x$. If $n < x < n + \frac{1}{2}$, then $n + \frac{1}{2} < x + \frac{1}{2} < n + 1$, and $\left[x + \frac{1}{2}\right] = n$. If $n + \frac{1}{2} < x < n$, then $n + 1 < x + \frac{1}{2} < n + \frac{3}{2}$, so $n + 1 = \left[x + \frac{1}{2}\right]$. If $x = n + \frac{1}{2}$, then $\left[x + \frac{1}{2}\right] = [n + 1] = n + 1$.

9. The argument is a small adjustment to Problem 8.

10. By Problem 6 above, there are exactly $[n/p]$ multiples of p that are less than or equal to n. Each of these will appear as a factor in $n!$. Of these, $\left[[n/p]/p\right]$ are multiples containing p itself as a factor; hence, these are multiples of p^2, where p appears as a factor at least twice. Now set $n/p = q + (r/p)$, where $0 \le r \le p - 1$. Then $n/p^2 = (q/p) + (r/p^2) = q_1 + (r_1/p) + (r/p^2)$, where $0 \le r_1 \le p - 1$. Now $0 \le (r_1/p) + (r/p^2) \le ((p - 1)/p) + ((p - 1)/p^2)) = 1 - (1/p) + (1/p) - (1/p^2) = 1 - (1/p^2) < 1$. Hence, $[n/p^2] = q_1 = \left[[n/p]/p\right]$. Continuing in this way, we see that $[n/p^3] = \left[[n/p^2]/p\right]$ of these are actually multiples of p^3, where p appears as a factor three times. Repeating, we will arrive at a highest power, p^k, beyond which

$[n/p^j] = 0$. Thus,

$$E(n,p) = \left[\frac{n}{p}\right] + \left[\frac{n}{p^2}\right] + \cdots + \left[\frac{n}{p^k}\right].$$

Section 2.2.

2. *Hint:* Substitute the expressions of (p_k, q_k) into the equation of the line to show that each one gives a solution. To show that all solutions are of this form, suppose that (p, q) is an arbitrary solution and write

$$ap + bq = n$$

$$ap_0 + bq_0 = n.$$

Subtract the two equations and use the fact that g.c.d.$(a, b) = 1$.

3. *Hint:* Use any of the equations in Problem 1.

4. This is clear from the drawing.

Section 2.3

1. There are three cases to examine: (1) P and Q both odd; (2) P (or Q) even and Q (or P) odd; (3) P and Q both even.
   Case 1: Then $P - 1$ and $Q - 1$ are both even, so $2 \mid (P-1)(Q-1)$. Since all factors of P and Q are odd, d must be odd. Hence, $d - 1$ is even and $2 \mid (d-1)$.
   Case 2: Then $Q - 1$ is even, so $2 \mid (P-1)(Q-1)$. Since 2 does not divide both P and Q, 2 does not divide d. Thus, $d - 1$ is even, and $2 \mid (d-1)$.
   Case 3: Then $P - 1$ and $Q - 1$ are both odd, so $[(P-1)(Q-1)]/2$ is a half-integer with odd numerator. Since $2 \mid P$ and $2 \mid Q$, $2 \mid d$. Hence, $(d-1)/2$ is also a half-integer with odd numerator. Thus

$$\frac{(P-1)(Q-1)}{2} + \frac{d-1}{2} = \frac{[(P-1)(Q-1)] + (d-1)}{2}.$$

   is an integer.

3. The proof of Theorem 2.1 applies with the following adjustment: Suppose that $P = dP_0$ and $Q = dQ_0$. Then the diagonal of the rectangle will contain the $(d-1)$ lattice points (P_0, Q_0), $(2P_0, 2Q_0), \ldots,$ $((d-1)P_0, (d-1)Q_0)$ on its interior. These points correspond to the summands for which $[n(Q/P)] = [jP_0(Q_0/P_0)] = jQ_0$, $j = 1, 2, \ldots, (d-1)$. The remaining summands count lattice points below the diagonal. By symmetry, doubling the sum gives the total number

of lattice points in the rectangle plus $d - 1$, since the diagonal points are then counted twice. Thus,

$$2 \sum_{n=1}^{P-1} \left[n \frac{Q}{P} \right] = (P-1)(Q-1) + (d-1).$$

4. In the rectangle $OABC$, consider the lower left quadrant determined by the points $(0,0)$, $(P/2,0)$, $(P/2,Q/2)$, and $(0,Q/2)$. Since $P/2$ and $Q/2$ are not integers, no lattice points lie on the upper or right boundary of the rectangle determined by these points. The diagonal $y = (Q/P)x$ divides this rectangle into two triangles, T_1 and T_2, with no lattice points on their boundaries. The number of points interior to T_1 is given by $\sum_{n=1}^{P'} [n(Q/P)]$, while the number of points interior to T_2 is given by $\sum_{n=1}^{Q'} [n(Q/P)]$. Hence, their sum is the number of points interior to the small rectangle, namely

$$P'Q' = \frac{P-1}{2} \frac{Q-1}{2}.$$

Section 3.2

2. $A = \frac{1}{2}ab$. The lower boundary contains the $(a+1)$ points $(0,0)$, $(1,0)$, $(2,0), \ldots, (a,0)$; the right boundary contains the b points $(a,1)$, $(a,2), \ldots, (a,b)$. Because a and b are relatively prime, the hypotenuse contains no additional lattice points, so $B = a + b + 1$. By Pick's Theorem, $I = A - \frac{1}{2}B + 1 = \frac{1}{2}ab - \frac{1}{2}(a+b+1) = \frac{1}{2}(a-1)(b-1)$.

3. $P_1P_2P_3P_4$ is formed by two triangles, T_1 (determined by P_0, P_4, and $(2,1)$) and T_2 determined by P_3, P_4, and $(2,1)$). Their combined areas have $A = \frac{9}{2}$. Interior lattice points are P_1, P_2, P_3, P_4, $(2,1)$, $(4,2)$, $(4,1)$, $(5,2)$, so $B = 8$. Hence, Pick's Theorem does not hold. The polygon is not simple because $\overline{P_1P_2}$ and $\overline{P_3P_4}$ share the point $(2,1)$, which is not a vertex.

4. By Pick's Theorem (or other means), the area of the outer figure is $\frac{41}{2}$, while that of the inner figure is 2. Hence, the doubly connected region has area $\frac{41}{2} - 2 = \frac{37}{2} = 18\frac{1}{2}$. Pick's Theorem would give $A = I + \frac{1}{2}B - 1 = 11 + \frac{1}{2}(15) - 1 = \frac{35}{2} = 17\frac{1}{2}$, so Pick's Theorem does not hold.

5. Let $A = I + \frac{1}{2}B - 1$ be the area of the outer polygon, and $A' = I' + \frac{1}{2}B' - 1$ be that of the inner. Then $A_1 = A - A' = (I - I') + \frac{1}{2}(B - B')$. However, $I_1 = I - (I' + B')$, while $B_1 = B + B'$. By Pick's Theorem, we would have $A_1 = I_1 + \frac{1}{2}B_1 - 1 = I - I' - B' + \frac{1}{2}B + \frac{1}{2}B' - 1 = (I - I') + \frac{1}{2}(B - B') - 1$, which is one less than the true area.

6. The vertices of the rhombus are the points $(\pm b, 0)$, $(0, \pm a)$. In addition to those boundary points, each side contains $(d-1)$ non-vertex lattice points. The area $A = 4\left(\frac{1}{2}ab\right) = 2ab$. By Pick's Theorem, $A = I + \frac{1}{2}B - 1$, so $2ab = I + \frac{1}{2}[4 + 4(d-1)] - 1 = I + 2d - 1$, and $I = 2ab - 2d + 1$.

Section 3.3

1. The statement of Lemma 3.1 will hold if $\sqrt{2}$ is replaced by $1 + \epsilon$, $\epsilon > 0$. This is because any horizontal line segment of length greater than 1 at lattice-point height must contain at least one lattice point. However, because the diagonal distance between lattice points is $\sqrt{2}$, no smaller number will give the covering theorem.

3 a. No.

   b. *Hint:* See Problem 4.

   c. *Hint:* See Problem 4.

Section 4.3

1. Suppose $n = 4k+3 = p_1^{a_1} p_2^{a_2} \cdots p_r^{a_r}$. If each prime $p_i = 4k_i+1$, then each $p_i^{a_i}$ is of that form, as is the product $p_1^{a_1} p_2^{a_2} \cdots p_r^{a_r}$. Furthermore, if $p_i = 4k_i + 3$, but the associated exponent a_i is even, then $p_i^{a_i} = 4m + 1$ for some m. Hence, some prime is of the form $4k+3$, and its associated exponent is odd. But by Theorem 4.2, n cannot be written as a sum of two squares, and Theorem 4.1 follows.

2. Let $z = a - bi$, $z' = c + di$. Then $|z|^2 = a^2 + b^2$ and $|z'|^2 = c^2 + d^2$. $|zz'|^2 = (|zz'|)^2 = (|z||z'|)^2 = |z|^2|z'|^2 = (a^2 + b^2)(c^2 + d^2)$. But $zz' = (a - bi)(c + di) = (ac + bd) + (ad - bc)i$. Hence, $|zz'|^2 = (ac + bd)^2 + (ad - bc)^2$.

Section 4.5

1. No. $10 = 4(1) + 6$ and $R(10) = R(2 \times 5) = 4(2 - 0) = 8$.

2. $n = 12k + 9 = 3(4k + 3)$. Since 3 does not divide $4k$, 3 does not divide $4k + 3$. This means that 3 appears only one time as a factor of n. By Theorem 4.2, $R(n) = 0$.

3. Divisors of 1225 are $1, 5, 25, 7, 35, 175, 49, 245, 1225$. $A = 6$, $B = 3$. $R(n) = 4(A - B) = 4(6 - 3) = 12$.

4. The calculation is long but straightforward:
   $T(1225) = 1 + 4\sum_{k=0}^{35}\left[\sqrt{1225 - k^2}\right] = 3853$.

Section 5.1

1 a. If (x, y) is in the strip, then $\alpha x - \frac{1}{2} < y < \alpha x + \frac{1}{2}$. Hence, $-\alpha x + \frac{1}{2} > -y > -\alpha x - \frac{1}{2}$, and so we have $\alpha(-x) - \frac{1}{2} < -y < \alpha(-x) + \frac{1}{2}$. This shows that $(-x, -y)$ is also in the strip.

b. $d = 1/\sqrt{1 + \alpha^2}$.

c. Suppose that $\alpha > 0$ and irrational. Then either $|\alpha - [\alpha]| < \frac{1}{2}$ or $|\alpha - ([\alpha] + 1)| < \frac{1}{2}$. Let $q = [\alpha]$ or $[\alpha + 1]$, depending on which inequality holds. Then $-\frac{1}{2} < q - \alpha < \frac{1}{2}$, so $\alpha - \frac{1}{2} < q < \alpha + \frac{1}{2}$, showing that $(1, q)$ lies in the strip bounded by the lines $y = \alpha x \pm \frac{1}{2}$. This is also the first lattice point in the first quadrant portion of the strip that a perpendicular to the bounding lines will encounter as it moves up the strip. (Any other point (p_1, q_1) must have $p_1 > 1$, hence $q_1 > q$.) The line $y = -(1/\alpha)x + (k/\alpha)$ passes through $(1, q)$ if $k = \alpha q + 1 = \alpha[\alpha] + 1$ or $\alpha[\alpha] + \alpha + 1$.

2 a. If (x, y) is in the region, then $-1 < 2\alpha x^2 - 2xy < 1$. But this is equivalent to $-1 < 2\alpha(-x)^2 - 2(-x)(-y) < 1$. Thus, $(-x, -y)$ lies in the region.

b. By the symmetry of S just shown, we need to consider only non-negative x. Solving for y, the inequality $-1 \leq 2\alpha x^2 - 2xy \leq 1$ becomes $\alpha x - (1/2x) \leq y \leq \alpha x + (1/2x)$. Consider the upper bound $\alpha x + (1/2x)$. Clearly, $y \to +\infty$ as $x \to 0^+$, and $y \to \alpha x$ above as $x \to +\infty$. Thus, $y = \alpha x + (1/2x)$, $x > 0$, is a branch of a hyperbola with asymptotes at $x = 0$ and $y = \alpha x$. The lower bound $y = \alpha x - (1/2x)$ has $y \to -\infty$ as $x \to 0^+$ and $y \to \alpha x$ from below as $x \to +\infty$. Hence, $y = \alpha x - (1/2x)$, $x > 0$, is a branch of the conjugate hyperbola with asymptotes $x = 0$ and $y = \alpha x$. Symmetry with respect to the origin gives the other branch in each case.

Section 5.2

1. By definition, a convex set that contains two given points must contain the line segment joining them.

3. Let M_1, M_2 be two convex sets and let $M_1 \cap M_2$ be their intersection. If A and B are points in $M_1 \cap M_2$, then A and B are both in M_1 and A and B are both in M_2. Hence, $\overline{AB} \subset M_1$, $\overline{AB} \subset M_2$, showing that $\overline{AB} \subset M_1 \cap M_2$.

Section 5.3

1. $A = 6$; $A' = \frac{3}{2}$; $s = 5$. $(n + 1)^2 A' > (n + 10)^2$ for $n \geq 40$.

2. Consider a disk of radius 2 centered at $(0, 0)$. Draw eight radii, each passing from $(0, 0)$ through one or two lattice points. (Radii on the axes will pass through two.) Cut eight strips inward from the boundary of the disk, each of width ϵ_1 and length $1 + \epsilon_2$, each centered on one of the eight radii. The resulting nonconvex figure has area $4\pi - 8(1 + \epsilon_2)\epsilon_1$, arbitrarily close to 4π, has central symmetry, and contains $(0, 0)$ but no other lattice point.

3 a. If $P = (x, y)$, then $P_1 = (x+n_1, y+m_1)$ and $P_2 = (x+n_2, y+m_2)$.
   Hence, $P_1 - P_2 = (n_1 - n_2, m_1 - m_2)$.

b. Since C' is convex, it contains the point $-P_1 = (-x-n_1, -y-m_1)$,
   as well as the line segment joint $-P_1$ to P_2. Hence, c' contains the
   midpoint between $-P_1$ and P_2, with coordinates $\big((n_2-n_1)/2, (m_2-m_1)/2\big)$. Thus, C contains the lattice point (n_2-n_1, m_2-m_1), which
   is not $(0, 0)$ since $P_1 \neq P_2$.

Section 6.2

1. $\sqrt{|\Delta|} = \sqrt{184259} \approx 429.254$. Sketch the parallelogram described,
   letting $k = 429 < \sqrt{|\Delta|}$. We see that the region contains the points
   $(0, 1)$ and $(0, -1)$. In fact, $201 < 429 < \sqrt{|\Delta|}$, and $400 < 429 < \sqrt{|\Delta|}$, so Minkowski's First Theorem is satisfied at $(0, 1)$ and $(0, -1)$.

Section 6.3

1. $\Delta = \pi - e \approx 0.4233$. $\sqrt{|D|} = \sqrt{2|\Delta|} \approx 0.92$. Looking at the
   parallelogram determined by

$$\xi + \eta = 2x - (\pi + e)y = \pm\sqrt{|D|}$$

$$\xi - \eta = \quad - (\pi - e)y = \pm\sqrt{|D|}$$

we see that $(3, 1)$ is in the interior. Testing, we get $|\xi| |\eta| = |2(3) - (\pi + e)(1)| \approx |6 - 5.86| |0.4233| = 0.059262$. Since $\frac{1}{2}|\Delta| = \frac{1}{2}|\pi - e| \approx 2.1165$, Minkowski's Second Theorem is satisfied at $(3, 1)$.

Section 7.1

1 a. Let $P = (x, y)$ be any point in the plane. Then $T(P) = (ax+by, cx+dy)$ is also a point in the plane.

a. and c. Let $P_1 = (x_1, y_1)$ and $P_2 = (x_2, y_2)$. The vector form of the
   equation of the line determined by these points is given by

$$P_t = P_1 + t(P_2 - P_1)$$
$$= \big(x_1 + t(x_2 - x_1), y_1 + t(y_2 - y_1)\big).$$

$T(P_i) = (ax_i + by_i, cx_i + dy_i)$, for $i = 1, 2$. The line determined by
$T(P_1)$ and $T(P_2)$ is given by

$$Q_t = T(P_1) + t\big(T(P_2) - T(P_1)\big)$$
$$= \big(ax_1 + by_1 + t(ax_2 + by_2 - ax_1 - by_1),$$
$$cx_1 + dy_1 + t(cx_2 + dy_2 - cx_1 - dy_1)\big).$$

Applying the transformation T to P_t, above, one sees that $T(P_t) = Q_t$.

b. The general conic is given by $Ax^2 + Bxy + Cy^2 + Dx + Ey + F = 0$.
    Letting $x = ax_1 + by_1$ and $y = cx_1 + dy_1$ and carrying out the required
    calculations to simplify the expression, we arrive at an equation of
    the same form in the new variables x_1 and y_1.

2 a. The new vertices are $(0,0), (1,1), (1,3),$ and $(0,2)$.

b. The old and new areas are equal.

3. $\det T - \det \begin{pmatrix} 1 & 1 \\ 1 & 2 \end{pmatrix} - 2 - 1 - 1$. New vertices are $(0,0), (10,10),$
    and $(20,30)$. New area = old area = 50.

4 b. $\det T = \det \begin{pmatrix} 2 & 3 \\ 4 & 6 \end{pmatrix} = 12 - 12 = 0$. Hence T^{-1} does not exist.

Section 7.3

1. If $ap + bq = ap' + bq'$ and $cp + dq = cp' + dq'$, then $a(p - p') + b(q - q') = 0$ and $c(p - p') + d(q - q') = 0$. These latter equations
    show that $(p - p', q - q')$ is a solution to the homogeneous linear
    system $ax + by = 0, cx + dy = 0$. However, since $\Delta = ad - bc \neq 0$,
    the system has a unique solution $(0,0)$. Hence, $p = p'$ and $q = q'$.

2. Choose any T with $\Delta = \det T = \pm 1$.

3. We may assume that any such parallelogram has one vertex at $(0,0)$.
    Let $P = (x,y)$ and $Q = (u,v)$. Then $R = (x + u, y + v)$. Let T be
    the transformation given by $T = \begin{pmatrix} x & u \\ y & v \end{pmatrix}$. Then $OPRQ$ is the image
    under T of the fundamental square given by $(0,0), (1,0), (1,1), (0,1)$
    with area 1. The area A of $OPRQ$ has $A = |\det T| = |xv - uy| \geq 1$.

Appendix I, Section I.1

1. $1/(a + bi) = [1/(a + bi)][(a - bi)/(a - bi)] = (1 - bi)/(a^2 + b^2)$.
    Since $a + bi$ is not a unit, the $a^2 + b^2 \neq 1$, and $(a + bi)^{-1}$ is not a
    Gaussian integer.

2. $a^2 + b^2 > 1^2 + 0^2 = 1$, since either (1) both a and b are nonzero
    integers or (2) one is zero and the other has absolute value greater
    than or equal to 2.

3. Since a and b are both integers, so are their squares, and so is $a^2 + b^2$.

Appendix I, Section I.2

1. Suppose that $\overline{q} = fh$, where neither f nor h is a unit. Then $\overline{\overline{q}} = q = \overline{fh} = \overline{f} \cdot \overline{h}$, where neither $\overline{f}$ nor $\overline{h}$ is a unit, contradicting the
    assumption that q is prime.

2. Clearly, $a + bi \neq (a - bi)(-1)$. If $a + bi = (a - bi)(i) = ai + b = b + ai$,
    then we must have $a = b$. But then $a + bi = a + ai = a(1 + i)$, a
    nontrivial factorization of the prime $q = a + bi$, unless $a = \pm 1$. An
    identical argument works for $u = -i$.

Bibliography

Blichfeldt, H. F. "A New Principle in the Geometry of Numbers with Some Applications." *Transactions of the American Mathematical Society* 15:3 (July 1914):227–35.

_____. *Finite Groups of Linear Homogeneous Transformations*. Part 2 of *Theory and Applications of Finite Groups* by G. A. Miller, H. F. Blichfeldt, and L. E. Dickson. New York: Wiley, 1916. Reprinted, New York: Dover, 1961.

_____. *Finite Collineation Groups*. Chicago: University of Chicago Press, 1917.

_____. "The Minimum Values of Positive Quadratic Forms in Six, Seven, and Eight Variables." *Mathematische Zeitschrift* 39 (1934):1–15.

Cassels, J. W. S. *Introduction to the Geometry of Numbers*. Classics of Mathematics Series. Corrected reprint of 1971 edition. Berlin: Springer, 1997.

Davenport, Harold. "The Geometry of Numbers." *Math. Gazette* 31 (1947): 206–10.

_____. *The Higher Arithmetic*. New York: Dover, 1983.

Dickson, L. E. *History of the Theory of Numbers, Vol. I: Divisibility and Primality*. Washington, D.C.: Carnegie Institute, 1919.

_____. *History of the Theory of Numbers, Vol. II: Diophantine Analysis*. Washington, D.C.: Carnegie Institute, 1920.

_____. *History of the Theory of Numbers, Vol. III: Quadratic and Higher Forms*. Washington, D.C.: Carnegie Institute, 1923.

Gauss, C. F. *Werke*. Göttingen: Gesellschaft der Wissenschaften, 1863–1933.

Grace, J. H. "The Four Square Theorem." *Journal of the London Mathematical Society* 2 (1927):3–8.

Grossman, Howard D. "Fun with Lattice Points." *Scripta Mathematica* 16 (1950):207–12.

Gruber, P. M., and C. G. Lekkerkerker. *Geometry of Numbers.* 2nd edition. Amsterdam and New York: North-Holland, 1987.

Hajós, G. "Ein neuer Beweis eines Satzes von Minkowski." *Acta Litt. Sci. (Szeged)* 6 (1934):224–5.

Hardy, G. H., and E. M. Wright. *An Introduction to the Theory of Numbers.* 5th ed. Oxford: Oxford University Press, 1983.

Hermite, Charles. "Lettres de Hermite à M. Jacobi." *J. reine angew. Math.* 40 (1850):261–315.

———. *Comptes Rendus Paris* 37 (1853).

———. *J. reine angew. Math.* 47 (1854):343–5, 364–8.

———. "Sur une extension donnée à la théorie des fractions continues par M. Tchebychev." *J. reine angew. Math.* 88 (1879):10–15.

———. *Oeuvres*, Vol. I. Paris: E. Picard, 1905.

———. *Oeuvres*, Vol III. Paris: Gauthier-Villars, 1912.

Hilbert, David, and S. Cohn-Vossen. *Geometry and the Imagination.* Translated by P. Nemenyi. New York: Chelsea, 1952.

Honsberger, Ross. *Ingenuity in Mathematics.* New Mathematical Library Series, Vol. 23. Washington, DC: MAA, 1970.

Hurwitz, A. "Über die angenäherte Darstellung der Irrationalzahlen durch rationale Brüche." *Mathematische Annalen* 39 (1891):279–84.

Koksma, J. F. *Diophantische Approximationen.* New York: Chelsea, 1936.

Korkine, A., and E. I. Zolotareff. "Sur les formes quadratiques positives quaternaires." *Mathematische Annalen* 5 (1872):581–3.

———. "Sur les formes quadratiques." *Mathematische Annalen* 6 (1873): 366–89.

———. "Sur les formes quadratiques positives." *Mathematische Annalen* 11 (1877):242–92.

Lyusternik, L. A. *Convex Figures and Polyhedra.* First Russian edition (1956) translated and adapted by Donald L. Barrett. Boston: D. C. Heath, 1966.

Minkowski, Hermann. "Über die positiven quadratischen Formen un über kettenbruchähnliche Algorithm." *J. reine agnew. Math.* 107 (1891):209–12.

———. *Ausgewahlte Arbeiten zur Zahlentheorie und zur Geometrie. Mit D. Hilbert's Gedachtnisre auf H. Minkowski (Göttingen, 1909). [Selected Papers on Number Theory and Geometry. With D. Hilbert's Commemorative Address in Honor of H. Minkowski.]* Teubner-Archiv zur Mathematik, Vol. 12. E. Kratzel and B. Weissbach, eds. Leipzig: Teubner, 1989.

――――. *Geometrie der Zahlen*. Bibliotheca Mathematic Teubneriana, Vol. 40. Leipzig: Teubner, 1910. First section of 240 pages appeared in 1896. Reprinted, New York and London: Johnson Reprint Corp., 1988.

――――. *Diophantische Approximationen: Eine Einfuhrung in die Zahlentheorie*. Reprinted, New York: Chelsea, 1957.

Mitchell, H. L., III. *Numerical Experiments on the Number of Lattice Points in the Circle*. Stanford, CA: Stanford University, Applied Mathematics and Statistics Labs, 1961.

Mordell, L. J. "On Some Arithmetical Results in the Geometry of Numbers." *Compositio Math*. 1 (1934):248–53.

Niven, Ivan. *Irrational Numbers*. Carus Mathematical Monographs, No. 11. Washington, DC: MAA, 1956.

――――. *Numbers: Rational and Irrational*. New Mathematical Library Series, Vol. 1. Washington, DC: MAA, 1961.

Niven, Ivan, and Herbert Zuckerman. "The Lattice Point Covering Theorem for Rectangles." *Mathematics Magazine* 42 (1969):85–86.

Olds, Carl D. *Continued Fractions*. New Mathematical Library Series, Vol. 9. Washington, DC: MAA, 1963.

Ore, Oystein. *Number Theory and Its History*. New York: McGraw-Hill, 1948. Reprinted with supplement, New York: Dover, 1988.

Schaaf, William. *Bibliography of Recreational Mathematics*, Vol. I. Reston, VA: National Council of Teachers of Mathematics, 1959; reprinted, 1973.

Scott, W. T. "Approximation to Real Irrationals by Certain Classes of Rational Fractions." *Bulletin of the American Mathematical Society* 46 (1940):124–9.

Sierpinski, W. *The Elementary Theory of Numbers*. 2nd edition. Andrzej Schinzel, ed. North-Holland Mathematical Library, Vol. 31. Amsterdam and New York: North-Holland; Warsaw: Polish Scientific Publishers, 1988.

Tchebychef, P. L. *Oeuvres de Tchebychef*. Translated into French by A. Markoff and N. Sonin. Reprinted, New York: Chelsea.

Uspensky, J. V., and M. A. Heaslet. *Elementary Number Theory*. New York: McGraw-Hill, 1939.

Index

$C(\sqrt{n})$, 45
$\Gamma(x)$, 105, 121
∞, 46
$\lim_{n \to \infty}$, 46
M-set, 65–67, 113
$n!$, 27
$N(n)$, 45
$R(n) = R(n = p^2 + q^2)$, 48, 56
$\mathbb{R}^{n+1}$, 83
$T(n)$, 50
$u \equiv v \pmod{m}$, 56
$[x]$, 14, 25–28
y-space, 74

absolute value of a complex number, 133
additive structure of plane, 133
admissible lattice, 145
affine transformation, 73–74
approximating irrationals, 105–107
arithmetic–geometric mean inequality, 80
associate primes, 135
axes, major and minor, 100; rotating, 113

Bertrand, J., 152
binary quadratic form, 97, 104
Blichfeldt, Hans Frederik, life of, 153–4; on approximating irrationals 124–9, 131; a packing theorem, 149; proof of Minkowski's Fundamental Theorem, 72–73, 117; on quadratic minima, 105, 121
Blichfeldt's Theorem, 113–6, 119–21; generalization of, 116
Bolzano-Weierstrass Theorem, 15
boundary of a polygon, 36
Bólyai, János, 141

center of symmetry, 66
checkerboard lattice, 148–9
cluster points, 15–16
common divisor, 55
complex numbers, 54, 133–4
congruence classes, 140–1
congruence notation, 56, 108, 140–1
congruent numbers, 140
conjugate of z, 134
consecutive sides, 35
contracting an M-set, 66
convex point set, 65
covering, of a lattice point, 114

critical determinant, 146
critical lattice, 146

Davenport, Harold, 108
de M'eziriac, Claude Bachet, 107
degree of approximation, 105
determinant Δ of a lattice 73–74, 85; of a linear transformation, 85
determinant, critical, 146
difference points, 116
dilating an M-set, 67
Diophantine approximations, simultaneous, 82
Diophantine equation, 28, 54
Diophantine inequalities, 63
Diophantus, 51, 54, 107
discriminant d, 98
doubly connected polygon, 37

Einstein, Albert, 152
Eisenstein, F., 152
ellipse, 23, 100–101
equivalent lattices, 89, 91–92, 95
equivalent primes, 135
Euclid's algorithm, 10, 136
Euler, Leonhard, 97, 107; on representations of n, 52, 54–56
expanding an M-set, 66
exterior of a polygon, 36

face-centered cubic (fcc) lattice, 148
Fermat, Pierre de, 52, 54, 107
Fibonacci sequence, 32
Fibonacci, Leonardo, 53
field, 140
four-dimensional sphere, 109
fractional part of x, 26
Frederick the Great, Prussian Academy of, 107
fundamental lattice L, 3–4, 88
fundamental parallelogram, 89–90
fundamental parallelopiped, 74
fundamental point-lattice $Lambda$, 3–4, 18, 23, 88, 90
Fundamental Theorem of Arithmetic, 8, 135
Fundamental Theorem of Complex Arithmetic, 136, 138–9

gamma function, 105, 121

Gauss, Carl Friedrich, 152; on lattice points in circles 4, 45–46, 48, 57–58; on packing 148; on quadratic forms, 97, 105

Gaussian integers, 133–4; factorization of, 134–5; unique factorization of, 138–9

Gaussian primes, 139–4

general affine transformation, 73–74

general lattice, 88

geometry of numbers, 3, 63, 75, 85,113

Girard, Albert, 52

greatest common divisor, 5, 7

greatest integer function, 14, 25–28

Harriot, Thomas, 148

Hermite, Charles, 63, 105, 108, 123, 152

hexagon, regular, 146

Hilbert, David, 151

Hurwitz, Adolf, 151

infimum, 146

integers, divisibility property, 7

integers, expressing, in standard form, 52

integers, Gaussian, 133–4; factorization of, 134–5; unique factorization of, 138–9

integers, relatively prime, 5, 95

integral part of x, 25

integral solutions, 28–31

intercepts, 30

interior of a polygon, 36

inverse transformation, 86–87

irrational numbers, 5; approximating, 80, 105, 123–31

Jacobi, Carl Gustav Jacob, 57, 58–59, 63

Jordan, Camille, 152

Kepler conjecture, 148

Kepler, Johannes, 148

Kirchhoff, R., 151

Korkine, A., 101, 104, 105, 149

Kronecker, Leopold, 151

Kummer, E., 151

Lagrange's Theorem, 107–10

Lagrange, Joseph Louis, 97, 107

lattice, admissible, 145; checkerboard, 148–9; critical, 146; face-centered cubic (fcc), 148; general, 88

lattice path, 32

lattice point, 3–4, 88; covering property, 38, 114; covering theorem, 38; visible, 95–96

lattice square, 72

lattice systems, 3, 4–10

lattices, equivalent, 89, 91–92, 95

least common multiple, 8

Legendre, Adrien Marie, 56, 97

Lie, Sophus, 153

linear transformation, 85, 87

Liouville's identity, 58

Liouville, Joseph, 58

lower bound of $|f(x, y)|$, 98

Mersenne, Marin, 54

minimum of $|f(x, y)|$, 98, 104

Minkowski, Hermann, 151–3; his geometric point of view, 3, 63–64, 75, 101, 109, 113; on quadratic minima, 105

Minkowski Theorem, A (for approximating irrationals), 123–9

Minkowski's First Theorem, 76–80

Minkowski's Fundamental Theorem, 67; applications of, 75–76, 83–84, 98, 100–101, 113, 125; proofs of 67–73, 117–9; in y-space, 74

Minkowski's General Theorem, 73, 109, 131

Minkowski's Second Theorem, 79–81

Minkowski's Third Theorem, 82

Mitchell, H. L., III, 46

multiplicative inverses, 134

multiplicative structure of the plane, 133

Niven, Ivan, 42

nontrivial factorization, 135

numbers, complex, 54, 133-4; congruent, 140; irrational, 5; rational 5, 8; real, 5

packing of circles, 146–7; of lattices, 145–6; of spheres, 145, 147–9

parallel displacement, 113

parallelogram, fundamental, 89–90; primitive, 89–90

parallelopiped, fundamental, 74

path, lattice point–free, 17–23

path of maximum width, 17–20; of width d, 18

Pick's Theorem, 36

Pick, Georg, 36

plane, structure of the, 133

point-lattice, construction of, 88–89; transformation of, 91–95

point-lattices, equivalent, 89

points, difference, 116

polygons, 35–36; doubly connected, 37

positive definite quadratic form, 98, 102, 104
prime Gaussian integer, 135
primitive parallelogram, 89–90
prism, 124
Pythagorean Theorem, 133, 137

quadratic form, binary, 97, 104; positive definite, 98, 102, 104
quadratic representation, 97–98
quadratic residues, 108

Raleigh, Sir Walter, 148
rational numbers, 5, 8; approximating irrationals by, 105–7
real numbers, 5
relatively prime Gaussian integers, 135
relatively prime integers, 5–6, 95
representation of an integer n, 48–53, 110; representation, quadratic, 97–98
representations of prime numbers, 54–56
rhombus, 37
ring, 134
rotating axes, 113

sides of a polygon, 35
simple polygon, 35
simultaneous Diophantine approximations, 82
slope formula, 11
slope, irrational, 5, 10–17; rational, 5, 6–10
Smith, H. J. S., 152

sphere, four-dimensional, 109
sphere packing, 145, 147–9
standard form for integers, 52
Steinhaus, Hugo, 36
symmetry about the origin, 22, 64
symmetry, center of, 66

Tchebychev, P. L., 123, 129
Thue, Alex, 54
transformation, affine, 73; inverse, 86; linear 85–87, 90–95; point-lattice, 90–95
translating, 113
translation of an M-set, 68

Unique Factorization Theorem, 52
unit area, 20, 95
unit square, 72, 90
units of a ring, 134

vectors, 74, 89
vertices, 35
visible points, 95–96
Voigt, W., 151
volume of a four-dimensional sphere, 109
von Helmholtz, H. L. F., 151

Weber, H., 151
Weierstrass, Karl, 151
Wilson's Theorem, 140

Zolotareff, E. I., 102, 104, 105, 149
Zuckerman, Herbert, 42

C. D. Olds was born in Wanganui, New Zealand, in 1912 and emigrated to the United States with his family at age 12. He won a scholarship to Stanford University where he received his B.A., M.A., and Ph.D. He taught mathematics at San Jose University from 1945 until his retirement in 1976. During his lifetime he published numerous papers in mathematical journals, was editor of *The Mathematical Log*, the official publication of the National High School and Junior College Club, and was the author of *Continued Fractions*, published in the New Mathematical Library series by the MAA. In 1973 he received the MAA's Chauvenet Prize, given to the author of an outstanding expository article on a mathematical topic for his paper "The Simple Continued Fraction Expansion of e," published in the *American Mathematical Monthly*. An active MAA member, he served as both President and Secretary-Treasurer of the Northern California Section of the MAA. Olds was a champion Breast Stroke swimmer at Stanford University and was also an excellent sailor, winning overall championships in his club races for two consecutive years. He died in 1979 shortly after having completed the manuscript for *The Geometry of Numbers*.

Anneli Lax, Editor of the New Mathematical Library from 1961 until 1999, died on September 24, 1999. Throughout a distinguished career, Lax worked tirelessly for the MAA; for nearly forty years she was a driving force in the MAA publications program. She was central to the publication of the New Mathematical Library (renamed the Anneli Lax New Mathematics Library in 1999), handling every aspect including acquisitions, copyediting, mathematical editing, layout and design, and typesetting.

Anneli Lax received the MAA's highest honor, the Yueh-Gin Gung and Dr. Charles Y. Hu Award, in 1995 for Distinguished Service to Mathematics. Professor Lax taught at New York University and was a pioneer in developing (in 1980) a combined course in expository writing and mathematical thinking. For many years before and after her retirement from NYU, she was active in helping to clarify and reform mathematical instruction at the school and college levels.

Giuliana Davidoff did her graduate work at the Courant Institute of Mathematical Sciences at New York University where she received the Ph.D. in 1984. While there she had the great good fortune to fall under the guiding intelligence of Anneli Lax. Together they worked on various undergraduate curriculum projects and wrote several articles, including

"Learning Mathematics," which was published in *Mathematics Tomorrow*. Her mathematical work is in number theory and automorphic forms. A member of the editorial board of the MAA Carus Mathematical Monograph series, Davidoff also serves on the editorial board of the MAA Dolciani Mathematical Expositions series. She is a Professor of Mathematics at Mount Holyoke College in South Hadley, Massachusetts.